T0330995

Intelligent Manufacturing and Industry 4.0

The use of intelligence in manufacturing has emerged as a fascinating subject for academics and businesses everywhere. This book focuses on various manufacturing operations and services which are provided to customers to achieve greater manufacturing flexibility, as well as widespread customization and improved quality with the help of advanced and smart technologies. It describes cyber-physical systems and the whole product life cycle along with a variety of smart sensors, adaptive decision models, high-end materials, smart devices, and data analytics.

Intelligent Manufacturing and Industry 4.0: Impact, Trends, and Opportunities focuses on Intelligent Manufacturing and the design of smart devices and products that meet the demand of Industry 4.0, manufacturing and cyber-physical systems, along with real-time data analytics for Intelligent Manufacturing. The usage of advanced smart and sensing technologies in Intelligent Manufacturing for healthcare solutions is discussed as well. Popular use cases and case studies related to Intelligent Manufacturing are addressed to provide a better understanding of this topic.

This publication is ideally designed for use by technology development practitioners, academicians, data scientists, industry professionals, researchers, and students interested in uncovering the latest innovations in the field of Intelligent Manufacturing.

Features:

- Presents cutting-edge manufacturing technologies and information to maximise product exchanges and production
- Discusses the improvement in service quality, product quality, and production effectiveness
- Conveys how a manufacturing company's competitiveness can increase if it can manage the turbulence and changes in the global market
- Presents how intelligence production is essential in Industry 4.0 and how Industry 4.0 offers greater manufacturing flexibility, as well as widespread customisation, improved quality, and increased productivity
- Covers the ways businesses handle the challenges of generating an increasing number of customised items with quick time to market and greater quality
- Includes popular use cases and case studies related to intelligent manufacturing to provide a better understanding of this discipline

Intelligent Manufacturing and Industrial Engineering

Series Editor: Ahmed A. Elngar, Beni-Suef Uni.
Mohamed Elhoseny, Mansoura University, Egypt

Machine Learning Adoption in Blockchain-Based Intelligent Manufacturing
Edited by Om Prakash Jena, Sabyasachi Pramanik, Ahmed A. Elngar

Integration of AI-Based Manufacturing and Industrial Engineering Systems with the Internet of Things
Edited by Pankaj Bhambri, Sita Rani, Valentina E. Balas and Ahmed A. Elngar

AI-Driven Digital Twin and Industry 4.0: A Conceptual Framework with Applications
Edited by Sita Rani, Pankaj Bhambri, Sachin Kumar, Piyush Kumar Pareek, and Ahmed A. Elngar

Technology Innovation Pillars for Industry 4.0: Challenges, Improvements, and Case Studies
Edited by Ahmed A. Elngar, N. Thillaiarasu, T. Saravanan, and Valentina Emilia Balas

Internet of Things and Big Data Analytics-Based Manufacturing
Edited by Arun Kumar Rana, Sudeshna Chakraborty, Pallavi Goel, Sumit Kumar Rana, and Ahmed A. Elngar

Industrial Internet of Things Security: Protecting AI-Enabled Engineering Systems in Cloud and Edge Environments
Edited by Sunil Kumar Chawla, Neha Sharma, Ahmed A. Elngar, Prasenjit Chatterjee, and P. Naga Srinivasu

Intelligent Manufacturing and Industry 4.0: Impact, Trends, and Opportunities
Edited by Alka Chaudhary, Vandana Sharma, and Ahmed Alkhayyat

For more information about this series, please visit: https://www.routledge.com/Mathematical-Engineering-Manufacturing-and-Management-Sciences/book-series/CRCIMIE

Intelligent Manufacturing and Industry 4.0

Impact, Trends, and Opportunities

Edited by
Alka Chaudhary, Vandana Sharma, and
Ahmed Alkhayyat

CRC Press
Taylor & Francis Group
Boca Raton London New York

CRC Press is an imprint of the
Taylor & Francis Group, an **informa** business

Designed cover image: Shutterstock

First edition published 2025
by CRC Press
2385 NW Executive Center Drive, Suite 320, Boca Raton FL 33431

and by CRC Press
4 Park Square, Milton Park, Abingdon, Oxon, OX14 4RN

CRC Press is an imprint of Taylor & Francis Group, LLC

ISBN: 978-1-032-62761-8 (hbk)
ISBN: 978-1-032-63073-1 (pbk)
ISBN: 978-1-032-63074-8 (ebk)

DOI: 10.1201/9781032630748

Typeset in Times
by Deanta Global Publishing Services, Chennai, India

Contents

Preface

Intelligent Manufacturing includes cutting-edge manufacturing technologies and information to maximise product exchanges and production. Over the whole product life cycle, a variety of smart sensors, adaptive decision models, high-end materials, smart devices, and data analytics can be helpful. There will be an improvement in service quality, product quality, and production effectiveness. A manufacturing company's competitiveness may increase if it can manage the turbulence and changes in the global market. The use of intelligence in manufacturing has emerged as a fascinating subject for academics and businesses everywhere. Intelligence production is essential in Industry 4.0. Industry 4.0 offers greater manufacturing flexibility, as well as widespread customisation, improved quality, and increased productivity. As a result, it helps businesses handle the challenges of generating an increasing number of customised items with quick time to market and greater quality.

The focus of this edited volume will be on Intelligent Manufacturing to design smart devices and products to meet the demand of Industry 4.0, manufacturing systems and cyber-physical systems, and real-time data analytics in Intelligent Manufacturing. The usage of advanced Smart and Sensing technologies in intelligent manufacturing for healthcare solutions is discussed in detail.

About the Editors

Alka Chaudhary received her Ph.D. in Computer Science and Engineering from Manipal University Jaipur, Rajasthan, India in 2016. She obtained her M.Tech in Computer Science and Engineering (with a Gold Medal) from Jagannath University Jaipur, Rajasthan, India and MCA (with Honors) from the Institute of Information Technology Ghaziabad, affiliated with Uttar Pradesh Technical University, India. She joined the Amity Institute of Information Technology (AIIT) at Amity University, Noida, Uttar Pradesh, India, in August 2019 and working as an assistant professor Grade III. She also served as the head of the IoT and Sensor Network Research Lab at AIIT. Her research interests encompass wireless network security, IoT, Artificial Intelligence, and soft computing techniques. She has published over 80 research papers in Scopus and SCIE-indexed international journals and conferences. She has filed 13 patents and 5 copyrights in AI and IoT. In recognition of her contributions, she was honored with the Young Scientist Award by the VDGOOD Professional Association in 2021. She has completed various certifications from CISCO NetAcad and has served as a Cisco Instructor. Currently, she is guiding four research scholars. She is an active member of professional bodies such as ACM, ACM-W, IAENG, and the Internet Society, and frequently reviews for several prominent journals.

Vandana Sharma is an associate professor at CHRIST (Deemed to be University), Delhi NCR, India. Dr. Sharma has served NAAC-accredited A++ institutions like Guru Gobind Singh Indraprastha University, Delhi and Galgotias University, Greater Noida Campus and Amity University Noida Campus, India. Dr. Sharma has 15+ years of teaching experience at the postgraduate level and has guided 100+ student dissertations. She is a senior member of IEEE, member of Women in Engineering Society and member of IEEE Consumer Technology Society popularly known as CTSoc. As a keen researcher, she has published 115+ research papers in SCI and Scopus Indexed international journals and conferences. Dr. Sharma has contributed voluntarily as a keynote speaker, session chair, reviewer and Technical Program Committee (TPC) member for reputed International Journals and IEEE Conferences and has presented her work across India and abroad. Currently, she is a post-doctoral fellow at Lincoln University College, Malaysia. Also, she is honorary adjunct professor and senior fellow of Scientific Innovation Research Group (SIRG) at Beni Suef University, Egypt and adjust Professor at Perdana University, Malaysia. Her primary areas of interest include Artificial Intelligence, Blockchain Technology and the Internet of Things (IoT).

Ahmed Alkhayyat received a B.Sc. degree in Electrical Engineering from A L Kufa University, Najaf, Iraq, in 2007, an M.Sc. degree from the Dehradun Institute of Technology, Dehradun, India, in 2010, and a Ph.D. from Cankaya University, Ankara, Turkey, in 2015. He contributed to organizing several IEEE conferences, workshops, and special sessions. He is currently a dean of International Relationships and manager of the world ranking at the Islamic University, in Najaf, Iraq. To serve his community, he acted as a reviewer for several journals and conferences. His research interests include IoT in the healthcare system, software-defined networking (SDN), network coding, cognitive radio, efficient-energy routing algorithms, and efficient-energy MAC protocol in cooperative wireless networks and wireless body area networks, as well as cross-layer designing for self-organized networks.

Contributors

Nidhi Agarwal
Lincoln University College
Selangor, Malaysia

Naeem Ahmad
Department of Computer Application
NIT Raipur
Raipur, India

Mary Cade T. Ambojia
Chief Nursing Officer & Infection
    Control Nurse Velazco Hospital
Cavite, Philippines

Misbah Anjum
Amity Institute of Information
    Technology
Amity University
Noida, Uttar Pradesh, India

Devershi Pallavi Bhatt
Manipal University Jaipur
Jaipur, India

Prachi Dahiya
Delhi Technological University
Delhi, India

Ritu Gautam
Computer Science & Applications
Amity University
Noida, Uttar Pradesh, India

Gautam Samblani
Manipal University
Jaipur, India

Shivam Goel
Amity University
Noida, Utter Pradesh, India

Shreya Kagrawal
Amity University Noida
Noida, Uttar Pradesh, India

Dr. Upinder Kaur
Department of Computer Science and
    Engineering
Akal University
Talwandi Sabo
Punjab, India

Prableen Kaur
Computer Science & Applications
AIMT
Ambala, India

Shivangi Mishra
Amity University
Noida, Uttar Pradesh, India

Mahima Shanker Pandey
Shardha University
Greater Noida, India

Seema Rani
Amity Institute of Information
    Technology
Noida, Uttar Pradesh, India

Shuchi Sethi
Amity Institute of Information
    Technology, (AIIT)
Noida, Uttar Pradesh, India

Aparna Sharma
Shardha University
Greater Noida, Uttar Pradesh, India

Manik Sharma
Computer Science & Application
DAV University
Jalandhar, India

Nidhi Sindhwani
Amity Institute of Information
    Technology
Noida, Uttar Pradesh, India

Umang Kant
Kite University Ghaziabad
Ghaziabad, Uttar Pradesh, India

Archana Singh
Global Institute of Information
    Technology
Greater Noida, Uttar Pradesh, India

Sarvesh Tanwar
Amity University
Noida, Uttar Pradesh, India

Ayush Thakur
Amity Institute of Information
    Technology
Amity University Noidae
Noida, Uttar Pradesh, India

1 Intelligent Manufacturing
Components, Challenges, and Opportunities

*Ritu Gautam, Prableen Kaur, Manik Sharma,
Alka Chaudhary, and Vandana Sharma*

1.1 INTRODUCTION

Intelligent Manufacturing (IM) emanates from the concepts and advancement of Artificial Intelligence (AI). Composed of both human specialists and intelligent machines, IM systems integrate human and machine intelligence. Intelligent systems can do intelligent tasks such as analysis, judgement, reasoning, ideation, and decision-making. Intelligent Manufacturing incorporates features of leading manufacturing techniques to achieve smart processes of manufacturing to meet the demands of dynamic international markets [1]. Across comprehensive supply chains of manufacturing, various small- and medium-sized industries, organisations as well as major corporations, IM makes all manual operations and flow of information available when and where they are required. Some cutting-edge technologies are needed for IM so that devices can adapt their behaviour to various circumstances according to their learning capacities and past experiences.

These technologies facilitate direct connectivity with industrial systems, which enables prompt problem-solving and adaptive decision-making [2]. Over the years, the field of Intelligent Manufacturing has evolved. Many relevant ideas have been circulated, and many are excited about the promising future of Industry 4.0. In truth, many modern manufacturers have significantly increased their level of automation (particularly in China). Thus, new generation technologies viz. cloud computing, AI, IoT, and big data are viewed as a means to enable IM. IM is also called Smart Manufacturing or Industry 4.0. Some key concepts of IM are:

Data: Intelligent Manufacturing relies heavily on big industrial data, particularly considering recent developments in IoT and cloud computing. Although data is essential to the analysis of manufacturing processes [3], there are still major obstacles in the way of current production. Many of the data are unstructured and have no relationship to the model, which reduces their significance for further big data analysis. Furthermore, accessing dark data can be difficult in many situations [4].

Artificial Intelligence: AI has impacted manufacturing on several levels in recent years. Genetic algorithms and deep neural networks optimise complicated tasks like resource allocation and production line scheduling at the workshop/factory level [5]. Researchers use convolutional neural networks for prediction to identify anomalous signals, and time series models such as long short-term memory for wear estimates during processing at the machine workpiece level [6]. AI is also used to investigate surface roughness, processing of deformation, and material temperature distribution at processing and material level. While AI tackles a variety of domain specific manufacturing problems, necessitates it requires the input of expert knowledge and depends upon huge high-quality datasets for training the model. AI used in isolation is unable to handle the larger issues involved in moving the manufacturing sector closer to intelligence [7,8].

Industrial Internet: When the Industrial Internet first emerged, it was thought to be similar with Industry 4.0, and peoples' hopes stemmed from the significant revolution brought about by the Internet. Manufacturing may become more intelligent through the convergence of cloud computing, IoT and big data, all of which could be facilitated by the Industrial Internet platform [9]. The idea of cloud manufacturing emerged from the Industrial Internet vision and was later understood to be an application for the general notion of Cyber Physical Systems (CPS) The industrial Internet is as essential to attaining Intelligent Manufacturing as big data is, and both depend on high-dimensional technologies to function at their best.

Modelling: The advancement of manufacturing depends on modelling technology, which is present in entire phases of the domain from design and production to testing and maintenance [10]. Present-day intelligent systems, propelled by various modelling techniques, generate isolated islands of intelligence at the prototype level. Figure 1.1 presents the general framework of Intelligent Manufacturing.

1.2   IM BASIC PARADIGMS

The term IM refers to a broad category of specialised subjects. The latest manufacturing technology and advanced AI are integrated into a new-generation IM. It passes via each stage of the creation, manufacturing, product, and service life cycles [11]. The idea also pertains to the integration and optimisation of related systems; its goal is to continuously improve the performance, quality, and service levels of businesses while consuming fewer resources. This will support the manufacturing industry's innovative, environmentally friendly, coordinated, open, and shared development. However, there are three basic IM paradigms shown below.

1.2.1   DIGITAL MANUFACTURING (DM)

First generation IM is also called DM. It is the initial and basic paradigm of IM [12]. The following are the characteristics of DM:

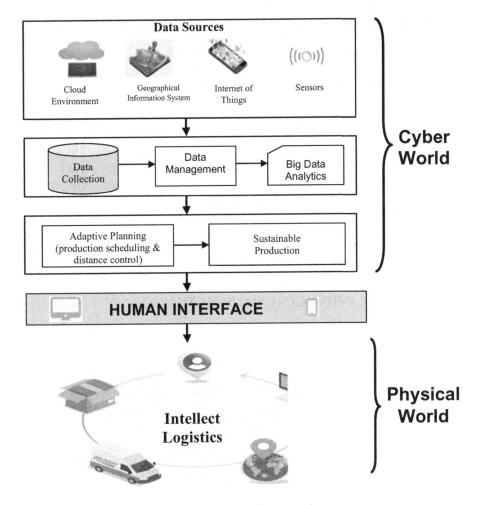

FIGURE 1.1   General framework of Intelligent Manufacturing.

- Digital technology applied on processes and products to form digital innovative products [13].
- Simulations and digital design are widely used for digital equipment.
- Optimisation of the production processes is accomplished.

1.2.2 Digital Network Manufacturing (DNM)

Another fundamental paradigm of Internet Manufacturing (IM) is called DNM. [14]. The following are the characteristics of DNM:

- Network and digital technologies are frequently applied at the product level. Through the network, products are connected, and joint and cooperative design as well as shared R&D are accomplished.

- End-to-end, horizontal, and vertical integration are finished at the manufacturing level, linking data flow of the entire system.
- Enterprises shift from product-centric production to user-centric production at the service level, where they connect and communicate with users via network platforms.

1.2.3 NEW-GENERATION IM

The latest fundamental paradigm of IM is known as digital-networked IM or new-generation IM. Manufacturing's intrinsic development pattern is reflected in its three fundamental paradigms. In a sense, the three fundamental paradigms represent the various stages of development of advanced manufacturing and information technologies since they each evolved in turn and had distinct qualities and major issues that needed to be resolved [15]. The three fundamental paradigms, however, are technologically inseparable and are updated through iterative processes, demonstrating the integrated development features of Intelligent Manufacturing.

1.3 NEED FOR INTELLIGENT MANUFACTURING

The need for IM arises from various challenges and demands in modern industrial landscapes [16]. Some of the points are discussed below:

Enhanced Efficiency and Productivity: Intelligent Manufacturing systems can automate repetitive tasks, optimise production processes, and minimise downtime. Due to these positive parameters, manufacturers can produce more goods in the least amount of time and can enhance productivity and efficiency.

Improvement in Quality: Intelligent Manufacturing facilitates real-time production data monitoring and analysis by employing advanced sensors and AI-driven technologies. Moreover, quality standards or product quality can be enhanced by timely identification of bugs in the product.

Reduction in Cost: Cost savings is one of the important needs of the Intelligent Manufacturing system. Labour cost reduction and minimum wastage of material can be achieved by optimisation and automation features of an Intelligent Manufacturing system. As a result, manufacturers can reduce overall costs and can make their business more competent.

Customization and Flexibility: Intelligent Manufacturing systems are flexible systems and can easily adapt to the changing requirements of the users, production schedules, and product specifications. Additionally, there is a faster response to consumer requests for individualised and customised items.

Predictiveness in Maintenance: IoT and AI technologies empower predictive maintenance, which assists manufacturers in foreseeing equipment breakdowns before they happen. Preventive scheduling of maintenance procedures can minimise unpredictive downtime and improve the lifespan of systems by recognising possible faults early on.

Optimised Supply Chain: IM can enhance interaction and communication in all areas of the supply chain. Manufacturers could improve supply chain efficiency, optimise inventories, and synchronise production schedules with suppliers through analytics and real-time data sharing.

Competition: To remain competitive in today's quickly changing market environment, producers need to constantly innovate and adapt to the latest technologies. IM systems can easily adapt the latest technologies to enhance operations, develop new products, and provide better products to customers with greater value.

Product Sustainability: Intelligent Manufacturing may facilitate sustainability by maximising resource use, cutting waste, and limiting environmental effects. By adopting more efficient processes and technologies, manufacturers can contribute to a more sustainable future while also meeting regulatory requirements and consumer expectations for eco-friendly practices.

Overall, Intelligent Manufacturing is essential for modern manufacturers looking to stay competitive, meet evolving customer demands, improve operational efficiency, and contribute to sustainable economic growth.

1.4 KEY COMPONENTS OF IM

The three key components of IM are digitalisation, networking, and intelligence. Figures 1.2 and 1.3 present components of IM and paradigms of IM, respectively [17]. Based on key components, some major components such as big data analytics, Machine Learning, IoT, advanced materials, robots, CPS, cybersecurity, cloud computing, and Digital Twins are also summarised. Table 1.1 and Figure 1.3 present the major components of IM.

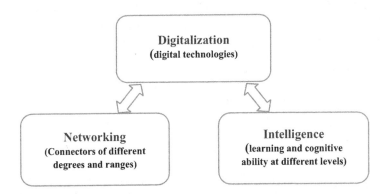

FIGURE 1.2 Core component of IM.

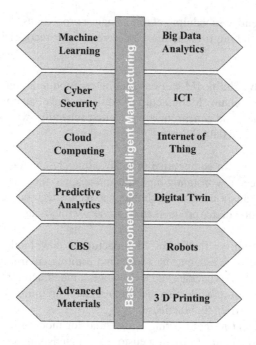

FIGURE 1.3   Some major key components of IM.

1.4.1   BIG DATA ANALYTICS

In today's IoT context, big data analytics is becoming essential. Identifying systematic patterns and making better decisions refers to the process of obtaining knowledge and information from huge data by locating clusters and correlations. Today, a vast amount of data is accessible from equipment, production, logistics, and user input throughout the manufacturing process [18]. Conventional analysts can't handle this kind of volume of data as these were either inaccessible or nonexistent in the old production setting. Big data analysis techniques like correlation and clustering, statistical modelling, and cognitive machine learning are among the new methods and schemes being explored. Using just pertinent and essential information from terabytes or more of datasets, big data analytics makes it possible to make the appropriate decision in manufacturing. This strategy will change the way industrial systems are controlled from reactive to proactive decision-making. Big data analytics is becoming most popular and significant for modern manufacturing systems due to its potential for manufacturing applications [19].

1.4.2   MACHINE LEARNING

Learning ability is one of the main traits of human intellect. By using data-driven algorithms, Machine Learning (ML) enables a computer to comprehend and learn about the internal workings of a physical system. ML techniques include data

TABLE 1.1

Characteristics of Key Components of IM

S. No	Technology/Key Component	Characteristics
1	Big Data Analysis and Analytics	Gather knowledge from enormous volumes of data to make informed choices.
2	Cloud Computing	Utilising cloud computing in products to increase their functionality and provide associated services.
3	CyberSecurity	Safety of Internet-connected systems from cyber-attacks.
4	CyberPhysical Systems (CPS)	Creating international business networks to bring the digital and physical worlds closer together.
5	Machine Learning	Machine Learning techniques facilitate production of actionable intelligence by analysing gathered data to enhance industrial productivity while maintaining a minimal impact on the necessary resources.
6	Predictive Analytics	Predictive Analytics is a branch of advanced analytics that forecasts future events using statistical modelling, machine learning and data mining, in conjunction with historical data.
7	Advanced Materials	Improved materials are required for the manufacture of modern goods.
8	3D Printing	A relatively new development, 3D printing offers a revolutionary manufacturing method that makes it possible to create systems and parts that are stronger and lighter.
9	Collaborative Robots	The production process has been accelerated by robot manufacturing.
10	Digital Twin	A new tool in intelligent manufacturing called the "digital twin" may assess the health of systems in real-time and anticipate system faults.
11	ICT	Software engineering, embedded systems, mobile technology, e-commerce, and computer communication networks are all made possible by ICT.
12	IoT	IoT refers to a networked and connected environment in where various objects have electronic sensors, actuators, or other digital devices implanted in them to enable data interchange and gathering.

mining, artificial neural networks (ANNs), and statistical pattern recognition algorithms [20].

1.4.3 CLOUD COMPUTING

An Internet-based computing service is offered via cloud computing, enabling software sharing and eliminating the need for users to install required software locally. Software as a Service (SaaS) is a common term for this activity. However, software sharing through the Internet is no longer sufficient for industrial system implantations. To establish a marketplace for software and knowledge sharing, it is also essential to share information and knowledge. "Platform as a Service" (PaaS) is the term for this technique [21]. PaaS for manufacturing applications are being developed and put into use.

1.4.4 IoT

Internet of Things (IoT) describes a networked and connected environment where different items are embedded with sensors, actuators, and other digital devices to facilitate data collection and exchange. IoT can generally provide enhanced connectivity for physical things, systems, and services, allowing data sharing and object-to-object communication. IoT can be used in a variety of industries to automate and control tasks, including robotic vacuums, lighting, heating, machining, and remote monitoring. Automatic identification, or auto-ID, is a crucial component of the Internet of Things and can be applied to create smart devices [22]. IoT offers IM systems a special and crucial base that may link every component of a manufacturing system to one another. In this approach, it is possible to greatly enhance both the efficiency of data gathering and the quality of the collected data. By 2025 , around 20.8 billion devices are expected to be fully connected and utilising RFID, according to reports [23]. The majority of industries will be impacted by this change, particularly the industrial sectors. Numerous items have been identified using RFID technology at distribution centres, manufacturing floors, logistics firms, warehouses, retail locations, and disposal/recycling phases.

1.4.5 Cyber Physical Systems (CPS)

CPS were first conceptualised as a cornerstone of IM. Cyber Physical Systems entails information sensing and analysis in physical space, digital machine creation in cyber-space, and generation of a CPS model of a machine tool on the cognitive layer [24]. Numerous industries such as aerospace, automotive, healthcare, and manufacturing have created and employed CPS devices. Many people consider this generation to be embedded systems. Also, CPS serve as the foundation for IoT deployment [25]. The development of sensing techniques and cloud computing led to the emergence of mobile CPS. Two essential requirements and obstacles for mobile CPS are security and reliability. Global and long-distance mobile CPS are still to come.

1.4.6 Cybersecurity

Network technology advancements in mobile network systems have made self-privacy and business security is crucial. This is true not just for information protection but also—and maybe more importantly—for the security of complex manufacturing systems [26]. Companies are creating and putting into place internal cybersecurity measures at an increasing rate. The US government is creating the necessary legislative protection and technological advancements to improve industrial cybersecurity. In order to exchange best practices and technology and enable the industry to successfully solve the security challenge, the US National Institute of Standards and

Technology has designed a cybersecurity framework. As many businesses get ready to deploy Industry 4.0, cybersecurity will continue to be their top priority [27].

1.4.7 ADVANCED MATERIALS

To produce contemporary goods, new and superior materials are desperately needed. Because of its less weight and superior material qualities, carbon fibre has been quickly embraced by the industry. Lately, there has been a notable advancement in the manufacture of carbon nanotubes. Significant demands also exist in high-tech fields, such as the need for novel battery materials and 3D printing materials. The ability to store energy, be lightweight, process information, have a smart memory, and other desirable qualities are sought in novel materials [28].

1.4.8 PREDICTIVE ANALYTICS

Nowadays, predictive analytics appears to be the most promising and successful use of big data technologies in industrial settings. Manufacturing organisations have come to the realisation that a vast quantity of data is accessible across their systems and is either underutilised or discarded. Predictive analytics could be the most promising way to address this shortcoming [29]. Many businesses are vying to create algorithms for learning and analysis to provide useful analytics that can forecast the state of machinery or other equipment in the future. This reduces equipment downtime and allows for more effective maintenance.

1.4.9 INFORMATION AND COMMUNICATIONS TECHNOLOGY

ICT stands for information and communication technology, which is expanded to include unified communications, telecom integration, and other technologies that may store, transfer, and alter data or information. ICT includes a broad spectrum of signal processing and computer science methods, including enterprise middleware, wireless systems, and audio-visual systems. It is essential in IM where production processes and decision-making greatly rely on the data, and it focuses on information transfer using a variety of electronic media, including standards for wired and wireless communication.

ICT has been proven to have a distinct effect on how a firm is organised, with improved ICT for workers and plant managers being linked to greater control and autonomy [30].

1.4.10 COLLABORATIVE ROBOTS

The ability of collaborative robots to operate alongside human operators offers a special advantage to manufacturing systems. This makes robot systems more flexible as well as smart to deal with complex and manufacturing situations [31]. Nowadays, robots are not seen as isolated devices that exist independent of human interactions. In order to increase automation and lower costs, more robots will be employed in the production process. To achieve this, collaborative robotic technology is being created and applied quickly.

1.4.11   3D PRINTING

The development of new materials and additive manufacturing technologies has advanced significantly in recent years. Processes for product design and development as well as production efficiency could be greatly enhanced by additive manufacturing. Growing numbers of industries will use 3D printing technology for their manufacturing as new materials become available and machine accuracy improves. It has been predicted that manufacturers will begin to use 3D printing extensively in 2025. In reality, a few sectors, including the medical and aerospace industries, have already started employing 3D printing technology to produce essential parts [32].

1.4.12   DIGITAL TWIN (DT)

A DT is a multi-level virtual representation and provides a comprehensive description of a real-world production process or system. In general, a DT is one of the integrated systems with the ability to compute, simulate, track, and manage system status and operations [33]. With the fast development of virtual and data acquisition technology, DT will become a key research direction of IM [34]. Zhou et al. [35] proposed a framework for a knowledge-driven DT.

1.5   CHALLENGES

Though Intelligent Manufacturing systems possess the capacity to manage the multiplicity of difficulties and intricacies that today's industries experience, there are still some challenges that come up during system implementation. The construction of advanced IM systems and the upgrading of existing industries with IM technology are thought to be fraught with security risks, an inadequate level of system integration, a lack of return on investment in new technology, and financial difficulties, among other dependent variables [36]. The following lists the issues that Intelligent Manufacturing systems face.

1.5.1   IM SECURITY ISSUES

An IM system is the production system that shares data with end-users across manufacturing units via a centralised network infrastructure.

It is set up specifically over the Internet and needs network connectivity for this reason. Data and information must be secure throughout the system at different points with worldwide unique identity and end-to-end encryption of data when sharing information over the Internet [37]. As a result, each network node needs to be secured against outside threats and improper usage of data. Ensuring the security of the entire system and complete process is the most crucial factor to be considered when creating networked systems such as Intelligent Manufacturing systems [38].

1.5.2   Integration of Systems

Integrating new technology equipment with existing equipment is a problem in the development of an Intelligent Manufacturing system [39]. Because new and old devices are incompatible, there are several problems with introducing smart manufacturing technology. Antiquated devices operating on antiquated communication protocols could employ a different protocol than modern devices. Connecting machines to other machines and establishing connections between systems also require a more robust communication infrastructure. Modern production systems need IPv6 connectivity to accommodate multiple linked devices simultaneously.

1.5.3   Interoperability

The capability of various systems to comprehend and utilise one another's functionalities on their own is known as interoperability [40]. This feature makes data and information interchangeable without requiring consideration of the hardware or software developer. Systematic interoperability is the application of models, principles, standards, and norms. Technical interoperability establishes the software related to the technical and ICT environments, together with its tools and platforms. If standards and communication techniques are not appropriately matched, interoperability may not be reached successfully.

The restrictions of the system's interoperability are determined by differences in hardware capabilities, operational frequency, communication mode, and communication bandwidth, among other factors [41,42].

1.5.4   Safety in Integrated Human-Robot Operations (Cobot)

A cobot is a specific kind of robot that can collaborate with humans in the workplace while providing new paradigms for human-machine interaction (HMI) safely and tangibly [43,44]. Human-robot cooperation is defined by the International Federation of Robotics as a robot's capacity to communicate and cooperate with humans to carry out specific activities in an industrial setting [45]. The main priority should be the occupational health and safety of individuals working at the site; hazardous conditions should be avoided, and proper occupational health and safety standards need to be maintained. The main goal of industrial robots and the CPS' implementation should be to reduce workplace hazards brought on by many causes such as noise, mechanical, vibration, materials, radiation, electrical, and work conditions.

1.5.5   Being Multilingual

Multilingual operations should be handled by smart manufacturing systems, which should be able to translate human-language instructions into machine-language instructions that tell the machine to perform the appropriate task [46].

1.5.6 Investment Return on Novel Technology

The financial analysis and return on investment are closely scrutinised when converting an old production system to a new sophisticated technology [47]. The extra costs involved, output losses during an upgrade, and the amount of time needed to recoup investment returns with income from the current system all have an impact on the adoption of newer technologies.

1.5.7 Advanced Technical Skills Gap

A 36% technical skills gap prevents sectors from benefiting from smart industrial investments. Industries must employ workers with digital dexterity—people who comprehend and work with digital tools and manufacturing processes to support, enhance productivity, and eliminate delays with technical skill gap recovery—to successfully implement and advance smart factories. Similar work on smart industrial setups is referred to in [48,49].

1.5.8 Inaccurate Change Management

Inaccurate change management protocols combined with inadequate project requirements present another difficulty in the deployment of smart factories. Clear requirement definition, project scope definition, and detailed explanation of change management protocols across factory operations departments are critical to the start and achievement of smart factory initiatives.

1.5.9 Inadequate Senior-Level Input

The start of a smart factory and its full implementation process entails all industry management and operations divisions. Large-scale implementation is therefore unavoidable, and for a successful implementation and business plan moving forward, input from all senior-level boards is needed. Given that the change impacts every department, senior staff members may be reluctant to contribute enough for the smart factory transformation to be successful.

1.5.10 Unrealistic Expectations

Management and industry stakeholders may have different ideas about what is expected of the deployment of smart factories. Without experience, determining a project's impact might be difficult. The imprecise comprehension and viewpoints of stakeholders and management might also make it difficult to plan resources realistically or anticipate an instantaneous digital transition. Management and stakeholders need to understand that digital transformation is a process rather than a one-time event.

1.5.11 Selecting the Wrong Technology Partner

The success of the implementation depends on selecting the appropriate technology partner to ensure a seamless adoption of smart factories. Be sure to evaluate a technology partner's product offers and support before choosing one. Make sure your technology partner has the operational and technical know-how to oversee a project that aligns with your business culture and create a smart factory.

1.5.12 The Lack of Clear Aims, Goals, and Explanations

To prevent challenges in smart manufacturing, it is crucial to make sure the foundations are in place before starting any project involving Intelligent Manufacturing. Those in charge of carrying out the project must have a comprehensive awareness of not just what is required, but also the objectives and benefits it will offer the company.

1.5.13 A Deficit of Understanding Concerning the Requirements

Establishing and documenting the project needs is crucial. Following that, discuss these requirements with the business team. This process is similar to that of the goals, objectives, and reasons.

1.6 OPPORTUNITIES FOR INTELLIGENT MANUFACTURING

Global adoption of IM in the manufacturing system presents a significant business opportunity. Intelligent Manufacturing is expected to be adopted by small- to medium-sized businesses. There is a lot of room for automation in the manufacturing, transportation, lodging, food services, and warehousing sectors. These industries have the most potential for automation, thus they have emerged as the focal point for research on the application of cutting-edge technology to boost output and efficiency. The applications of smart manufacturing in the field of agriculture are referred to in [50].

Intelligent Manufacturing can help your business in sales and marketing by helping you comprehend markets and anticipate and adjust to customer preferences. Intelligent Manufacturing can assist with demand forecasting, inventory optimisation, and supplier monitoring of supply chain optimisation. Supply chain companies have long utilised analytics for inventory management and forecasting, but in the Internet of Things era, when nearly everything is tracked, more real-time skills are needed. 5G networks have the potential to revolutionise manufacturing. Industrial data can be used at scale because of 5G's capacity to accommodate tens of thousands of endpoints at high-connection densities. The core of programs employing excellent manufacturing, documentation, and security practices and Quality by Design (QbD) is the ability to comprehend and demonstrate that processes are in control. Standardization, automation, and oversight of QbD and GxP projects can be facilitated by Intelligent Manufacturing to comply with regulations. Even the most sophisticated firms face challenges when it comes to proving to authorities that their procedures are understood and under control.

Digital firms need to implement Intelligent Manufacturing since basic automation is not required to keep up with the market and Industry 4.0. Manufacturers need to use field- and customer-centric analytics if they want to survive the digital disruption brought about by Industry 4.0 and the Internet of Things. Figure 1.4 presents applications of IM applications.

1.7   SUMMARY

IM is the future of the industry and society. With the expeditious development of advanced technologies such as big data, robotics, AI, and IoT, Intelligent Manufacturing systems are ushering in miraculous opportunities and permeating deeply into different areas of society. These advanced manufacturing systems have replaced traditional manufacturing systems and revolutionised the industry. Due to the amalgamation of cutting-edge technologies with manufacturing industries, Intelligent Manufacturing systems can forecast the demands of manufacturers more accurately and achieve better planning of production with effective supply chain management. Early alert systems in Intelligent Manufacturing can promptly find and resolve problems, thereby enhancing production efficiency and contributing to the growth of the manufacturing industry. Additionally, smart design, smart monitoring, smart machines, smart scheduling, and smart control are the key benefits of Intelligent Manufacturing.

Although there have been many encouraging opportunities, Intelligent Manufacturing still encounters a number of problems that must be overcome before it can be successfully implemented widely. Nevertheless, due to uncertainty, manufacturing industries will continue to face significant difficulties such as interoperability, integration of components, data privacy, cybersecurity and upskilling of the manufacturers etc. In addition, the collaboration of humans and robots for flexible automation is another major challenge for Intelligent Manufacturing systems. In the future, deep learning techniques can be used for defect-free manufacturing. Moreover, deep learning can assist in the development of robot intelligence to an extent where they can support human operators while offering much enhanced context awareness for complete human safety.

In conclusion, while Intelligent Manufacturing brings forth significant benefits and opportunities, addressing associated challenges is crucial for its successful implementation and realization of its full potential. With strategic investments in technology, infrastructure, and human capital, companies can go through these challenges and harness the transformative power of Intelligent Manufacturing to drive sustainable growth and competitiveness in the evolving industrial landscape.

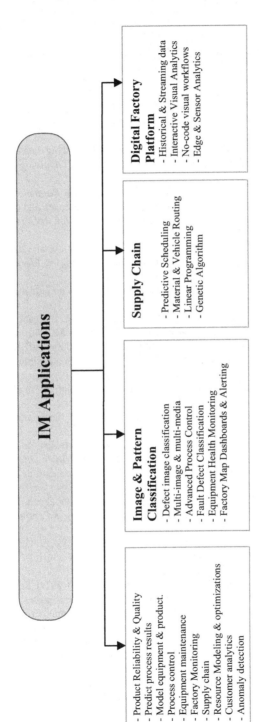

FIGURE 1.4   Applications of IM.

REFERENCES

1. Zhong, R. Y., Xu, X., Klotz, E., & Newman, S. T. (2017). Intelligent manufacturing in the context of industry 4.0: a review. *Engineering*, 3(5), 616–630.
2. Zhou, J., Li, P., Zhou, Y., Wang, B., Zang, J., & Meng, L. (2018). Toward new-generation intelligent manufacturing. *Engineering*, 4(1), 11–20.
3. Liang, S., Rajora, M., Liu, X., Yue, C., Zou, P., & Wang, L. (2018). Intelligent manufacturing systems: a review. *International Journal of Mechanical Engineering and Robotics Research*, 7(3), 324–330.
4. Li, C., Chen, Y., & Shang, Y. (2022). A review of industrial big data for decision making in intelligent manufacturing. *Engineering Science and Technology, an International Journal*, 29, 101021.
5. Pan, Y. (2016). Heading toward Artificial Intelligence 2.0. *Engineering*, 2(4), 409–413.
6. Zhuang, Y. T., Wu, F., Chen, C., & Pan, Y. (2017). Challenges and opportunities: from big data to knowledge in AI 2.0. *Frontiers of Information Technology & Electronic Engineering*, 18(1), 3–14.
7. Li, W., Wu, W., Wang, H., Cheng, X., Chen, H., Zhou, Z., et al. (2017). Crowd intelligence in the AI 2.0 era. *Frontiers of Information Technology & Electronic Engineering*, 18(1), 15–43.
8. Wang, L. (2019). From intelligence science to intelligent manufacturing. *Engineering*, 5(4), 615–618.
9. Lin, S. W., Miller, B., Durand, J., Joshi, R., Didier, P., Chigani, A., et al. (2015). *Industrial Internet Reference Architecture* [Internet]. Milford: Industrial Internet Consortium (IIC). Available from: https://www.iiconsortium.org/IIRA-1.7.htm.
10. Zhang, L., Zhou, L., Ren, L., & Laili, Y. (2019). Modeling and simulation in intelligent manufacturing. *Computers in Industry*, 112, 103123.
11. Zhou, J. (2015). Intelligent manufacturing—main direction of "made in China 2025". *China Mechanical Engineering*, 26(17), 2273–2284.
12. Zhou, J. (2013). Digitalization and intelligentization of the manufacturing industry. *Advances in Manufacturing*, 1(1), 1–7.
13. Brown, R. G. (2000). Driving digital manufacturing to reality. In J. A. Joines, R. R. Barton, K. Kang, and P. A. Fishwick PA, (Eds.), *2000 Winter Simulation Conference Proceedings* (pp. 224–248). Orlando, FL: Piscataway: Institute of Electrical and Electronics Engineers, Inc.
14. Westkämper, E. (2007). Digital Manufacturing in the global era. In P. F. Cunha and P. G. Maropoulos (Eds.), *Digital Enterprise Technology: Perspectives and Future Challenges* (pp. 3–14). Boston, MA: Springer US.
15. Jiyuan, Z. A. N. G., Baicun, W. A. N. G., Liu, M. E. N. G., & Yuan, Z. (2018). Brief analysis on three basic paradigms of intelligent manufacturing. *Strategic Study of Chinese Academy of Engineering*, 20(4), 13–18.
16. Wang, B., Tao, F., Fang, X., Liu, C., Liu, Y., & Freiheit, T. (2021). Smart manufacturing and intelligent manufacturing: A comparative review. *Engineering*, 7(6), 738–757.
17. Zhang, H., Nian, P., & Chen, Y. (2017, July). Intelligent manufacturing: The core leads industrial change in the future. In *2017 4th International Conference on Industrial Economics System and Industrial Security Engineering (IEIS)* (pp. 1–5). IEEE.
18. Wang, J., Xu, C., Zhang, J., & Zhong, R. (2022). Big data analytics for intelligent manufacturing systems: A review. *Journal of Manufacturing Systems*, 62, 738–752.
19. Nagorny, K., Lima-Monteiro, P., Barata, J., & Colombo, A. W. (2017). Big data analysis in smart manufacturing: A review. *International Journal of Communications, Network and System Sciences*, 10(3), 31–58.

20. Acosta, P. C., Terán, H. C., Arteaga, O., & Terán, M. B. (2020). Machine learning in intelligent manufacturing system for optimization of production costs and overall effectiveness of equipment in fabrication models. In *Journal of Physics: Conference Series* (Vol. 1432, No. 1, p. 012085). IOP Publishing.
21. Simeone, A., Caggiano, A., Boun, L., & Deng, B. (2019). Intelligent cloud manufacturing platform for efficient resource sharing in smart manufacturing networks. *Procedia CIRP*, 79, 233–238.
22. Haghnegahdar, L., Joshi, S. S., & Dahotre, N. B. (2022). From IoT-based cloud manufacturing approach to intelligent additive manufacturing: Industrial Internet of things— An overview. *The International Journal of Advanced Manufacturing Technology*, 119(3), 1461–1478.
23. Wang, K. S. (2014). Intelligent and integrated RFID (II-RFID) system for improving traceability in manufacturing. *Advances in Manufacturing*, 2, 106–120.
24. Dafflon, B., Moalla, N., & Ouzrout, Y. (2021). The challenges, approaches, and used techniques of CPS for manufacturing in Industry 4.0: a literature review. *The International Journal of Advanced Manufacturing Technology*, 113, 2395–2412.
25. Yao, X., Zhou, J., Lin, Y., Li, Y., Yu, H., & Liu, Y. (2019). Smart manufacturing based on cyber-physical systems and beyond. *Journal of Intelligent Manufacturing*, 30, 2805–2817.
26. Wu, D., Ren, A., Zhang, W., Fan, F., Liu, P., Fu, X., & Terpenny, J. (2018). Cybersecurity for digital manufacturing. *Journal of Manufacturing Systems*, 48, 3–12.
27. Ren, A., Wu, D., Zhang, W., Terpenny, J., & Liu, P. (2017). Cyber security in smart manufacturing: Survey and challenges. In *IIE Annual Conference. Proceedings* (pp. 716–721). Institute of Industrial and Systems Engineers (IISE).
28. Chen, Y. (2017). Integrated and intelligent manufacturing: Perspectives and enablers. *Engineering*, 3(5), 588–595.
29. Lee, J., Singh, J., Azamfar, M., & Pandhare, V. (2020). Industrial AI and predictive analytics for smart manufacturing systems. In *Smart Manufacturing* (pp. 213–244). Elsevier.
30. Kumar, K., Zindani, D., Davim, J. P., Kumar, K., Zindani, D., & Davim, J. P. (2019). *Intelligent manufacturing. Industry 4.0: Developments towards the Fourth Industrial Revolution* (pp. 1–17). Cham: Springer.
31. Sherwani, F., Asad, M. M., & Ibrahim, B. S. K. K. (2020, March). Collaborative robots and industrial revolution 4.0 (ir 4.0). In *2020 International Conference on Emerging Trends in Smart Technologies (ICETST)* (pp. 1–5). IEEE.
32. Chen, T., & Lin, Y. C. (2017). Feasibility evaluation and optimization of a smart manufacturing system based on 3D printing: A review. *International Journal of Intelligent Systems*, 32(4), 394–413.
33. He, B., & Bai, K. J. (2021). Digital twin-based sustainable intelligent manufacturing: A review. *Advances in Manufacturing*, 9(1), 1–21.
34. Zheng, Y., Yang, S., & Cheng, H. (2019). An application framework of digital twin and its case study. *Journal of Ambient Intelligence and Humanized Computing*, 10(3), 1141–1153.
35. Zhou, G., Zhang, C., Li, Z., Ding, K., & Wang, C. (2020). Knowledge-driven digital twin manufacturing cell towards intelligent manufacturing. *International Journal of Production Research*, 58(4), 1034–1051.
36. Phuyal, S., Bista, D., & Bista, R. (2020). Challenges, opportunities and future directions of smart manufacturing: A state of art review. *Sustainable Futures*, 2, 100023.
37. Tuptuk, N., & Hailes, S. (2018). Security of smart manufacturing systems. *Journal of Manufacturing Systems*, 47, 93–106.

38. Thoben, K.-D., Wiesner, S., & Wuest, T. (2017). "Industrie 4.0" and smart manufac-turing-a review of research issues and application examples. *International Journal of Automation Technology*, 11(1), 4–16.

39. Oztemel, E. (2010). Intelligent manufacturing systems. In L. Benyoucef and B. Grabot (Eds.), *Artificial Intelligence Techniques for Networked Manufacturing Enterprises Management. Springer Series in Advanced Manufacturing*. London: Springer. https://doi.org/10.1007/978-1-84996-119-6_1

40. Rao, M., & Luxhoj, J. T. (1991). Integration framework for intelligent manufacturing processes. *Journal of Intelligent Manufacturing*, 2, 43–52. https://doi.org/10.1007/BF01471335

41. Chen, D., Doumeingts, G., & Vernadat, F. (2008). Architectures for enterprise inte-gration and interoperability: Past, present and future. *Computers in Industry*, 59(7), 647–659.

42. Ray, S. R., & Jones, A. T. (2006). Manufacturing interoperability. *Journal of Intelligent Manufacturing*, 17, 681–688 (2006). https://doi.org/10.1007/s10845-006-0037-x

43. Tibaut, A., Rebolj, D., & Nekrep Perc, M. (2016). Interoperability requirements for auto-mated manufacturing systems in construction. *Journal of Intelligent Manufacturing*, 27, 251–262. https://doi.org/10.1007/s10845-013-0862-7

44. Gualtieri, L., Palomba, I., Wehrle, E. J., & Vidoni, R. (2020). The opportunities and challenges of SME manufacturing automation: Safety and ergonomics in human–robot collaboration. In *Industry 4.0 for SMEs* (pp. 105–144). Springer.

45. ISO/TS 15066robots and robotic devices–collaborative robots. (2016). International Organization for Standardization. *Technical Report, 1,1-30*.

46. Litzenberger, G. (2019). IFR Publishes Collaborative Industrial Robot Definition and Estimates Supply. *International Federation of Robotics. IFR Secretariat Blog. Zugriff am, 24*, 2020. .

47. Uhlemann, J., Costa, R., & Charpentier, J.-C. (2020). Product design and engineer-ing—past, present, future trends in teaching, research and practices: Academic and industry points of view. *Current Opinion in Chemical Engineering*, 27, 10–21.

48. Singh, A., Ali, A., Balusamy, B., & Sharma, V. (2023). Chapter 12 - Potential applica-tions of digital twin technology in virtual factory. In R. K. Dhanaraj, A. K. Bashir, V. Rajasekar, B. Balusamy, and P. Malik (Eds.), *Digital Twin for Smart Manufacturing* (pp. 221–241). Academic Press. ISBN 9780323992053. https://doi.org/10.1016/B978-0-323-99205-3.00011-0

49. Borkar, K. K., Aljrees, T., Pandey, S. K., Kumar, A., Singh, M. K., Sinha, A., Singh, K. U., & Sharma, V. (2023). Stability analysis and navigational techniques of wheeled mobile robot: A review. *Processes*, 11, 3302. https://doi.org/10.3390/pr11123302

50. Raj, A., Sharma, V., Rani, S., Shanu, A. K., Alkhayyat, A., & Singh, R. D. (2023). Modern farming using IoT-enabled sensors for the improvement of crop selection. In *2023 4th International Conference on Intelligent Engineering and Management (ICIEM)* (pp. 1–7). London, United Kingdom. https://doi.org/10.1109/ICIEM59379 .2023.10167225.

2 Navigating the Future of Intelligent and Smart Manufacturing
A Comprehensive Bibliometric Analysis (2012–2024)

Upinder Kaur and Vandana Sharma

2.1 INTRODUCTION

The manufacturing industry has undergone a remarkable transformation in the past decade, thanks to the rise of intelligent manufacturing and smart manufacturing. Combining advanced manufacturing and information technology leads to incredibly efficient, automated, and data-driven production processes. Assessing the wide scope of studies in this area becomes essential for gaining insights into its trajectory followed by predicting future improvements during this paradigm transition.

Additionally, we reviewed academic works from the Web of Science database in order to provide an in-depth evaluation of research published on intelligent and smart manufacturing from 2012 to 2024. We also decided upon 1046 peer-reviewed papers for this research to demonstrate the progress of the field over time. The aforementioned research papers were chosen from the initial pool of 1131 that we screened. An examination of the available research illustrates the depth and interdependence of the field. The clusters of research incorporate a wide range of subjects from academia, such as convolutional neural networks (CNN), technologies, and the Internet of Things (IoT).

In addition to delineating the present state of research, a fundamental aim of this investigation is to discern recurring trends in terms of research emphasis, cooperation, and innovation. We will discover the forces propelling the field's development and get insight into its most impactful works through this examination. As a result, updates and trends in smart manufacturing may be disseminated to stakeholders, encompassing industry practitioners and researchers.

Furthermore, the research investigates the significance of prominent scholarly contributors, the citation-based valuation of the worth of their work, and the international cooperation that is the foundation of this complex field of study. By functioning as a comprehensive account of the progress made thus far and as a guiding principle for the future as novel technologies and methodologies emerge, this analysis ensures that forthcoming investigations into intelligent and smart manufacturing are guided by informed judgement and strategically targeted approaches.

2.2 LITERATURE REVIEW

The confluence of the diverse trends in emerging technology breakthroughs with the wider scope of augmenting intelligence, more flexibility, and high reliability in manufacturing processes is at the core of defining intelligent and smart manufacturing. By synthesising the extant literature, this review emphasises significant advancements, theoretical frameworks, and empirical investigations in the field. Intelligent manufacturing is the buzzword in the AI era. The technologies have undergone a huge transformation and large-scale integration. The domain reveals that in-depth consideration is required for the unification of operational technology and its large computational capacity. Jones and Richey (2000) explored the potential of these technologies to advance production efficiency and product quality was highlighted [1]. The emphasised role of IoT, edge computing, cloud-based solutions, big data, global supply chain generation, and new transformations in Industry 4.0. The emerging importance on strengthening cyber-physical systems in establishing a new manufacturing paradigm with the rise of Industry 4.0 [2]. This integration was further expedited by Industry 4.0. Thus, it highlights those how global technologies helps in exhibits the transformative impact of emerging industry. The blend of digital and physical world, helps in enhancing the productivity, quality, flexibility with interconnected devices. The inclusion of cyber security make the infrastructure more resilient and protect the same against the potential threats and ensures the more secure and sustainable manufacturing technologies.

The five-tier smart factory paradigm, delineated by Lee et al. (2013) [3], provides a strategic progression from connection to cognition. This conceptual framework helps in putting forth and comprehending the workings of intelligent manufacturing systems. Similarly, Zhou et al. (2015)[4] presented a structure for intelligent manufacturing systems with a focus on the integration of computing, networking, and sophisticated sensing capabilities.

The fundamental support in intelligent and smart manufacturing is devoted to the challenges supported while reviewing the existing research. In Patel and Cassou's (2015) study [5], they emphasised the key importance of the Internet of Things (IoT) in facilitating data collection and analysis during production cases. Meanwhile, research is underway on the intersection of machine learning (ML) and artificial intelligence (AI) in the fields of predictive maintenance and quality control, as presented by O. Pisacane [6], which covers the efficacy in substantially diminishing periods of inactivity and enhancing the standard of products [6].

Researchers have learned a lot about the effects and uses of intelligent manufacturing technology via empirical studies [7]. In Cronin (2019), they concluded that embracing intelligent manufacturing translated into higher capacity utilisation and minimised costs related to operations [8]. Another study that highlighted the potential of AI to improve manufacturing precision and efficiency was Liu et al. [9]. The research consisted of the implementation of convolutional neural networks (CNN) to recognise abnormalities.

Nevertheless, the mainstreaming of intelligent manufacturing procedures faces limitations. Significant shortcomings that have been discovered include labor force competence, data security issues, and the failure to let go of antiquated technologies [10]. Green et al. (2022) presented the possibility of intelligent and smart manufacturing with the interoperability of cybersecurity measures, workforce skill development, and the promotion of interoperability standards [11]. As demonstrated by this literature review, intelligent and smart manufacturing research progresses from conceptual frameworks to technological implementation and empirical validation. A future study on the issue must solve hurdles and investigate new advances as it progresses.

2.3 METHODOLOGY

This chapter aims to extensively assess the present position of research in the fields of intelligence and smart manufacturing through a comprehensive bibliometric examination. We analyse 1046 publications with peer review published from 2012 to 2024 by employing data gathered from the Web of Science databases. The aforementioned bibliometric analytic tool, an R program designed for conducting comprehensive bibliometric research, forms the foundation of our methodological approach. Employing this application, you may quantify and visualise the structure, dynamics, and trends of the dataset. The result provides invaluable information about the progress of research on intelligent and smart manufacturing.

In this chapter the initial dataset entailed a keyword search inside the Web of Science database, employing phrases with "smart manufacturing" and "intelligent manufacturing." Consequently, an entire collection of 1131 articles was created according to the search criteria, with articles published between 2012 and 2024. A combined total of 1046 papers went forward for further analysis after undergoing a stringent screening procedure that focused on ensuring their level of quality and relevance.

2.3.1 BIBLIOMETRIC ANALYSIS

The bibliometric tool is used to generate the many dimensional results and do further analysis of those results. Finding and assessing separate research clusters in the intelligent and smart manufacturing sector was the primary goal of the investigation.

2.3.1.1 Quantitative Characteristics of the Cluster

The total number of publications in the top cluster is denoted by N. The metrics used for cluster analysis are cohesiveness metrics: k is the average degree of its publication, d is the density for top clusters where $d = 2k/(N-1)$, $<w_{in}>$ is the weighted density, Q_i is the inner modularity calculated by splitting the top cluster in sub- partition,

Corpus	N	$<N_{ref}>$	k	d	$<w_{in}>*10^3$	Q_i	q	h_{ref}	$nr_{10}/nr_5/nr_2$	$<PY>$	$<A>_{refs}$	$<N_{cit}>$	h
All in BC	1046	65.19	53.85	0.103	0.637	0.349	-	-	-	2021.70	5.95	14.59	59
Cluster 3	219	53.85	15.76	0.145	2.043	0.376	0.048	10	1/4/48	2021.72	5.09	13.63	28
Cluster 1	196	65.78	29.67	0.304	3.860	0.313	0.058	11	3/15/133	2021.73	6.77	15.28	29
Cluster 10	175	95.23	15.77	0.181	2.474	0.467	0.051	10	1/15/183	2021.88	5.40	15.30	26
Cluster 2	141	65.79	37.04	0.529	6.819	0.246	0.056	10	4/19/175	2021.42	5.60	19.81	26
Cluster 4	104	57.91	39.44	0.766	15.863	0.117	0.083	13	23/59/174	2022.00	5.46	15.74	21
Cluster 6	61	60.13	23.18	0.773	9.800	0.177	0.020	5	3/11/109	2021.54	6.52	11.97	13
Cluster 5	57	49.46	9.68	0.346	7.942	0.456	0.015	6	12/77/216	2021.44	7.39	9.72	13
Cluster 7	53	37.32	7.13	0.274	5.685	0.383	0.011	6	6/26/114	2021.68	7.45	4.92	9
Cluster 8	19	104.95	4.95	0.550	11.975	0.000	0.003	4	124/1847/1847	2021.63	9.01	15.42	8
Cluster 9	14	65.57	5.43	0.835	16.869	0.000	0.002	3	61/839/839	2021.57	5.37	20.29	7

FIGURE 2.1 A comprehensive summary of the classification of every research paper based on the specified clusters.

and q is the module of within the sub-partition. The estimate how our cluster can condense the available references and exhibit the h-index using h_{ref} and nr_{10}, nr_5, nr_2 where nr_x indicates the available references cited by at least x% of publication within the top cluster. We also exhibit the average year of publication within the top cluster $<PY>$, and the average age of references by $<A>$refs. The estimated number of citations per publication is represented by $<N_{cit}>$, and h shows that h publications have been cited at least h times. An overview of the categorization of each research article according to the designated clusters is depicted in Figure 2.1.

The quantitative characteristics of these clusters were assessed using a set of cohesiveness metrics, as follows:

- N (total number of publications): This metric represents the total number of articles within the top cluster, providing a measure of the cluster's size and scope.
- k (average degree): The average degree of publications within a cluster, indicating the average number of connections each publication has with others in the cluster.
- d (density): Calculated as $d = 2k/(N - 1)$, this metric measures the connectedness or cohesiveness of the cluster, with higher values indicating a more tightly-knit cluster.
- <win> (weighted density): Represents the weighted density within the cluster, accounting for the strength of connections between publications.
- Qi (inner modularity): The inner modularity is computed by dividing the top cluster into sub-partitions, providing insights into the cluster's internal structure.
- q (modularity of within sub-partition): This metric estimates the coherence within sub-partitions of the top cluster, offering a view of the cluster's internal consistency.
- H-index metrics: The h-index (href) and the nr10, nr5, nr2 metrics, where nrx represents the number of references cited by at least x% of publications within the top cluster. These metrics gauge the cluster's impact and the density of influential publications.
- <PY> (average year of publication): Indicates the average publication year within the top cluster, shedding light on the cluster's currency and evolution over time.

- <A>refs (average age of references): The average age of references used within the cluster, providing an estimate of the foundational literature's recency and relevance.
- <Nett> (average number of citations per publication): Represents the average citation count per publication, offering an indicator of the cluster's academic impact.
- h (h-index): The h-index for the cluster, reflecting the number of publications (h) that have been cited at least h times, underscoring the cluster's scholarly influence.

2.3.2 INTERPRETATION AND REPORTING

A thorough assessment of the intelligent and smart manufacturing research landscape is made possible by the bibliometric analysis offered by the biblio tool. This study not only finds the most dynamic research domains but also discerns significant trends, influential works, and patterns of collaboration by evaluating the quantitative attributes of selected clusters. The unveiling of the primary findings of this analysis indicates how crucial it is in comprehending the progression of the field, the lasting impact of pivotal performances, and the trajectory of subsequent academic endeavours. The techniques stipulated in this study are used to conduct bibliometric assessments in the field of intelligent and smart manufacturing research, delivering a solid framework for the analysis. Through the use of a focused approach to gathering and analysing data, along with the bibliometric analysis tool, we offer a comprehensive overview of the development and current state of the field. This offers an expanded awareness of its complicated and exciting possibilities for subsequent studies.

2.4 DISCUSSION AND RESULTS ANALYSIS

2.4.1 OVERVIEW OF FINDINGS

2.4.1.1 Dataset Overview

A thorough review was done of 1046 articles published in the period between 2012 and 2024. Articles were extracted from the Web of Science database. The articles covered a broad spectrum of contributions to smart manufacturing. It seems that there has been an uninterrupted increase in the number of publications over the years, with particularly significant increases in recent years. This would indicate that there is increasing interest and activity in the field.

2.4.1.2 Cluster Identification

Through our bibliometric inspection of the dataset, we succeeded in finding ten unique clusters by applying co-citation analysis, especially the keyword's co-occurrence analysis. By bringing together themes, technological advances, and methods, these clusters can be viewed as unified groups of articles. Clusters vary in number of residents, coupled with certain clusters experiencing a higher concentration of people than different people, highlighting the diverse nature of research in the field of intelligent manufacturing.

2.4.2 CLUSTER ANALYSIS

2.4.2.1 Cluster Characteristics

There are distinctive characteristics associated with each cluster based on its size, cohesion, and thematic focus. The largest cluster (IoT) contains 219 articles, characterised by high interconnectedness ($k = 30.47$) and moderate density ($d = 0.278$), indicating a cohesive community exploring IoT's uses in manufacturing. As contrasted with Cluster 9 (Remote Maintenance), Cluster 9 has a smaller number of articles but a high density ($d = 0.636$), which indicates a tightly-knit cluster focusing on a particular aspect of smart manufacturing. During analysis the most cited publications (according to Web of Science) and most representative publications (in term of in-degree d_{in} measuring the number of publications in the cluster that are linked with it) among all publications in the cluster. The most cited and representative authors are represented in cluster and for each author, we display the number N_a of publications they have authored in that cluster, the sum TC_a of their number of citations (according to Web of Science), and the sum k_a of their in-degree.

2.4.2.2 Cluster 1 (CNN—Convolutional Neural Networks)

Cluster 1, with 196 articles, demonstrates considerable engagement with CNN applications in the industrial sector. An extensive foundation in research is shown by the average number of references (Nref>), but a higher average degree (k) indicates robust interconnectedness within this cluster, signifying fruitful interactions and debates. By utilising measures such as the h-index and quality index (Qi), it is possible to ascertain the relevance and recentness of this study, as well as identify significant advancements and developing patterns in the use of CNN in the industrial sector. The articles within this cluster could go into the utilisation of CNN for the purpose of conducting visual inspections, evaluating quality, and overseeing industrial operations in real-time. Through a comprehensive examination of the titles, keywords, and references, one could have been able to deduce the progression of CNN from rudimentary image processing to intricate undertakings including anomaly detection, predictive maintenance, and control systems. Prominent publications and writers within this cluster would serve as indicators of those who are spearheading the use of deep learning within the domain of manufacturing. The publications that receive the highest number of citations are important pieces that established the foundation for using CNN in industrial environments and showcased the technology's ability to enhance precision and productivity. Figure 2.2 illustrates the specific outcomes for Cluster 1.

2.4.2.3 Cluster 2 (Technologies)

This cluster contains 141 articles that cover a variety of technologies related to intelligent manufacturing. Its bibliometric indicators, such as references and citations, reflect a wide range of technological influences. As indicated by data such as the h-index and average citations, the picture aids in understanding the cluster's maturity, research breadth, and influence. The analysis would include robots, automation, data analytics, and artificial intelligence, all of which are developing technologies

FIGURE 2.2 Cluster 1 bibliometric analysis report.

impacting the industrial sector. A bibliometric study will disclose how these technologies interact to provide highly responsive, flexible, and efficient industrial environments. Frequently, the most referenced publications propose basic frameworks or offer revolutionary systems that have significantly impacted research and practise. The most referenced books and prolific authors will demonstrate the most disruptive technology. Bibliometric data may reflect long-term patterns in research priority areas, such as the transition from basic automation to more integrated, intelligent systems. Furthermore, this cluster would highlight the important journals and conferences where these technologies are discussed the most. Figure 2.3 shows the exact outcomes for Cluster 2.

2.4.2.4 Cluster 3 (IoT—Internet of Things)

This cluster, which contains 219 articles, highlights the significance of IoT in manufacturing. Bibliometric metrics—including the average number of references and the density—ascertain the depth and coherence of a research document, denoted by (d). Significant publications and developments in this domain have impacted the IoT's function in intelligent manufacturing. This cluster may comprise scholarly articles

Cluster 2 ("TECHNOLOGIES"). This cluster contains $N = 141$ publications.

Keywords	f(%)	σ
INTERNET	19.86	3.42
MANAGEMENT	14.18	3.43
BIG DATA	11.35	2.54
FRAMEWORK	11.35	0.13
FUTURE	11.35	3.73
TECHNOLOGIES	11.35	5.66
CHALLENGES	10.64	2.35
MODEL	9.93	0.84
DESIGN	8.51	-1.57
IMPLEMENTATION	7.09	3.81
SYSTEM	7.09	-0.59
PERFORMANCE	6.38	1.18
CYBER-PHYSICAL SYSTEMS	5.67	1.88
SMART	5.67	0.69
SUPPLY CHAIN	4.96	3.18
THINGS	4.96	-0.44
ARCHITECTURE	4.26	-0.48
INDUSTRY 4.0	4.26	-0.59
OPTIMIZATION	4.26	-1.00
BARRIERS	3.55	3.65
Title Words	f(%)	σ
INDUSTRY	40.43	10.18
40	36.88	11.04
MANUFACTURING	29.08	-0.32
SMART	20.57	-2.23
INTELLIGENT	17.02	-1.55
SYSTEM	12.06	0.49
SYSTEMS	12.06	1.13
APPROACH	8.51	1.76
FRAMEWORK	8.51	1.88
MODEL	8.51	2.01
Journal	f(%)	σ
APPL SCI-BASEL	6.38	2.25
J INTELL MANUF	4.96	2.73
IEEE ACCESS	4.26	0.94
SENSORS-BASEL	3.55	0.51
ENG APPL ARTIF INTEL	2.84	2.26
PROCEDIA COMPUT SCI	2.84	2.26
PROCEDIA MANUF	2.84	3.36
COMPUT IND	2.13	1.58
COMPUT IND ENG	2.13	0.72
INT J ADV MANUF TECH	2.13	-0.20

Institution	f(%)	σ
DEPT COMP SCI	6.38	1.31
DEPT PROD ENGN	5.67	4.27
DEPT MECH ENGN	4.96	-1.29
FAC MECH ENGN	4.96	2.87
DEPT ELECT & COMP ENGN	3.55	1.59
SCH ECON & MANAGEMENT	3.55	1.99
SCH MECH ENGN	3.55	-0.69
UNIV BELGRADE	3.55	4.42
DEPT IND ENGN	2.84	0.27
IND ENGN DEPT	2.84	3.35
UNIV SAO PAULO	2.84	4.28
ZHEJIANG UNIV	2.84	0.67
BEIHANG UNIV	2.13	0.50
BRANCH UNIV TRIER	2.13	4.29
BRISBANE	2.13	3.54
DEPT IND & SYST ENGN	2.13	0.32
DEPT IND ENGN & MANAGEMENT	2.13	1.57
FAC ENGN	2.13	-0.00
GERMAN RES CTR ARTIFICIAL INTELLIGENCE DFKI	2.13	4.29
HUAZHONG UNIV SCI & TECHNOL	2.13	0.71
Country	f(%)	σ
Peoples R China	20.57	-4.90
India	9.93	0.56
Taiwan	6.38	0.51
UK	6.38	-0.73
Brazil	5.67	2.55
Germany	5.67	0.44
Italy	5.67	0.60
Poland	5.67	3.20
USA	5.67	-2.12
South Korea	4.26	0.22
Author	f(%)	σ
Bergmann R	2.13	4.30
Klein P	2.13	4.30
Majstorovic VD	2.13	4.30
Malburg L	2.13	4.30
Abonyi J	1.42	2.66
Chen BT	1.42	3.51
Cohen Y	1.42	2.66
Durakbasa NM	1.42	3.51
Gonzalez-Jimenez J	1.42	3.51
Khan MA	1.42	2.13

Reference	f(%)	σ
Lee Jay, 2015, Manufacturing Letters, 3, 18	34.75	15.06
Lasi H, 2014, BUS INFORM SYST ENG+, 6, 239	23.40	12.48
Xu LD, 2018, INT J PROD RES, 56, 2941	13.48	8.13
Frank AG, 2019, INT J PROD ECON, 210, 15	10.64	6.96
Oztemel E, 2020, J INTELL MANUF, 31, 127	9.22	6.00
Chen BT, 2018, IEEE ACCESS, 6, 6505	7.80	3.89
Bahrin MAK, 2016, J TEKNOL, 78, 137	7.09	6.94
Lee J, 2014, PROC CIRP, 16, 3	7.09	4.75
Liao YX, 2017, INT J PROD RES, 55, 3609	7.09	6.57
Monostori L, 2014, PROC CIRP, 17, 9	7.09	6.57
Vaidya S, 2018, PROCEDIA MANUF, 20, 233	7.09	5.04
Dalenogare LS, 2018, INT J PROD ECON, 204, 383	6.38	5.48
Wang SY, 2016, COMPUT NETW, 101, 158	6.38	2.72
Ghobakhloo M, 2018, J MANUF TECHNOL MANA, 29, 910	5.67	6.03
Kagermann H, 2013, Final report of the Industries, 4, 0	5.67	6.03
Monostori L, 2016, CIRP ANN-MANUF TECHN, 65, 621	5.67	2.89
Schumacher A, 2016, PROC CIRP, 52, 161	5.67	5.00
Xu X, 2021, J MANUF SYST, 61, 530	5.67	3.03
Zhong RY, 2017, ENGINEERING-PRC, 3, 616	5.67	0.47
Kusiak A, 2018, INT J PROD RES, 56, 508	4.96	0.13
RefJournal	f(%)	σ
INT J PROD RES	47.52	4.56
PROC CIRP	41.84	4.79
IEEE ACCESS	40.43	0.44
Manufacturing Letters	37.59	11.07
COMPUT IND	36.88	4.70
IEEE T IND INFORM	35.46	1.00
INT J ADV MANUF TECH	34.75	1.20
J MANUF SYST	33.33	1.05
SENSORS-BASEL	33.33	1.57
COMPUT IND ENG	32.62	4.11
Subject	f(%)	σ
Computer Science, Artificial Intelligence	19.86	3.38
Engineering, Manufacturing	19.15	0.48
Engineering, Electrical & Electronic	18.44	-0.17
Engineering, Industrial	12.77	-0.23
Computer Science, Information Systems	12.06	-0.45
Computer Science, Interdisciplinary Applications	12.06	0.52
Engineering, Multidisciplinary	11.35	1.44
Computer Science, Theory & Methods	9.93	2.19
Automation & Control Systems	9.22	-0.50
Management	9.22	5.31

Cluster 2 ("TECHNOLOGIES"). Most cited publications (according to Web Of Science) and most representative publications (in terms of in-degree d_{in}, measuring the number of publications in the cluster that are linked with it) among all publications in the cluster.

Most Cited publications

d_{in}	Times Cited	Publication Ref	Publication Title
95	806	Oztemel E, 2020, J INTELL MANUF(31)	Literature review of Industry 4.0 and related technologies
49	108	Ahmed I, 2022, IEEE T IND INFORM(18)	From Artificial Intelligence to Explainable Artificial Intelligence in Industry 4.0: A Survey on What, How, and Where
39	89	Kalsoom T, 2020, SENSORS-BASEL(20)	Advances in Sensor Technologies in the Era of Smart Factory and Industry 4.0
82	81	Silvestri L, 2020, COMPUT IND(123)	Maintenance transformation through Industry 4.0 technologies: A systematic literature review
82	80	Ruiz-Sarmiento JR, 2020, ENG APPL ARTIF INTEL(87)	A predictive model for the maintenance of industrial machinery in the context of industry 4.0
67	78	Ghobakhloo M, 2021, J CLEAN PROD(295)	Industry 4.0 and opportunities for energy sustainability
58	75	Rajput S, 2020, J CLEAN PROD(277)	Industry 4.0 Model for circular economy and cleaner production
2	74	Mbunge E, 2021, HUM BEHAV EMERG TECH(3)	A critical review of emerging technologies for tackling COVID-19 pandemic
37	71	Javaid M, 2020, J IND INTEGR MANAG(5)	Critical Components of Industry 5.0 Towards a Successful Adoption in the Field of Manufacturing
77	68	Shi Z, 2020, SYST RES BEHAV SCI(37)	Smart factory in Industry 4.0
23	55	[illegible], 2020, TECHNOL FORECAST SOC(156)	Smart Products value creation in SMEs innovation ecosystems
44	50	[illegible], 2020, TOTAL QUAL MANAG BUS(31)	Quality managers and their future technological expectations related to Industry 4.0
39	50	Rahman A, 2022, CLUSTER COMPUT(25)	SDN-IoT empowered intelligent framework for industry 4.0 applications during COVID-19 pandemic
17	43	Hernandez-de-Menendez M, 2020, INT J INTERACT DES M(14)	Competencies for Industry 4.0

Most Representative Publications

d_{in}	Times Cited	Publication Ref	Publication Title
95	806	Oztemel E, 2020, J INTELL MANUF(31)	Literature review of Industry 4.0 and related technologies
92	30	Nakayama RS, 2020, COMPUT IND ENG(144)	Towards I4.0: A comprehensive analysis of evolution from I3.0
90	24	Kumar S, 2022, NEURAL COMPUT APPL(35)	Towards design and implementation of Industry 4.0 for food manufacturing
84	42	Jain V, 2021, INT J PRODUCT PERFOR(70)	Modelling the enablers of industry 4.0 in the Indian manufacturing industry
82	81	Silvestri L, 2020, COMPUT IND(123)	Maintenance transformation through Industry 4.0 technologies: A systematic literature review
83	80	Ruiz-Sarmiento JR, 2020, ENG APPL ARTIF INTEL(87)	A predictive model for the maintenance of industrial machinery in the context of industry 4.0
82	22	[illegible], 2021, J MODEL MANAG(16)	Sustainable industry 4.0: an exploratory study for uncovering the drivers for integration
82	17	Ryalat M, 2023, APPL SCI-BASEL(13)	Design of a Smart Factory Based on Cyber-Physical Systems and Internet of Things towards Industry 4.0
82	8	Sun X, 2022, ENVIRON SCI POLLUT R(29)	Towards the smart and sustainable transformation of Reverse Logistics 4.0: a conceptualization and research...
79	10	Fakhri AB, 2021, CMC-COMPUT MATER CON(66)	Industry 4.0: Architecture and Equipment Revolution
79	6	[illegible], 2024, ENG APPL ARTIF INTEL(127)	[illegible], a general Deep Learning architecture for Predictive Maintenance within the industry 4.0 paradigm
77	68	Shi Z, 2020, SYST RES BEHAV SCI(37)	Smart factory in Industry 4.0
77	9	[illegible], 2022, APPL SCI-BASEL(12)	A Review of Deep Reinforcement Learning Approaches for Smart Manufacturing in Industry 4.0 and 5.0 Framework
74	32	Ghosh AK, 2021, J IND INF INTEGR(24)	Developing sensor signal-based digital twins for intelligent machine tools

Cluster 2 ("TECHNOLOGIES"). Most cited and representative authors. For each author, we display the number N_a of publications their authored in that cluster, the sum TC_a of their number of citations (according to Web Of Science), and the sum k_a of their in-degree.

Most Cited Authors

Author	N_a	TC_a	k_a
Oztemel E	1	806	95
Gursev S	1	806	95
Ahmed I	1	108	48
Jeon G	1	108	48
Piccialli F	1	108	48
Kalsoom T	1	89	39
Ramzan N	1	89	39
Ahmed S	1	89	39
Ur-Rehman M	1	89	39
Silvestri L	1	81	82
Forcina A	1	81	82
Introna V	1	81	82
Santolamazza A	1	81	82
Cesarotti V	1	81	82
Ruiz-Sarmiento JR	2	80	161
Monroy J	1	80	82
Moreno FA	1	80	82
Galindo C	1	80	82
Bonelo JM	1	80	82

Most Representative Authors

Author	N_a	TC_a	k_a
Kim J	3	27	171
Ruiz-Sarmiento JR	2	80	161
Gonzalez-Jimenez J	2	80	161
Malburg L	3	21	121
Klein P	3	21	121
Bergmann R	3	21	121
Wang LH	2	68	115
Zheng P	2	36	112
Chen BT	2	8	103
Oztemel E	1	806	95
Gursev S	1	806	95
Nakayama RS	1	30	92
Spina MD	1	30	92
Silva JR	1	30	92
Konur S	1	24	90
Lan Y	1	24	90
Thakker D	1	24	90
Morkyani G	1	24	90
Polovina N	1	24	90

FIGURE 2.3 Cluster 2 bibliometric analysis report.

that explore the ways in which sensors, actuators, and networked machines might be utilised to streamline communication among various manufacturing system components. The cluster may facilitate the identification of new trends in the Internet of Things (IoT), including but not limited to process optimisation, supply chain management, and real-time data collection. Additionally, it may shed light on prominent research institutes and nations. The transformation of the Internet of Things' function from peripheral connectivity to incorporation into core systems may be discerned through an analysis of citation patterns and keyword trends. Additionally, IoT research hotspots can be found by the study of this cluster. Figure 2.4 depicts the precise results pertaining to Cluster 3.

2.4.2.5 Cluster 4 (Services/Applications)

This cluster of 104 articles focuses on manufacturing services and applications, including distinctive measurements such as weighted in-degree (ωin). In addition to identifying major service models and applications influencing the manufacturing sector, bibliometric data may reveal pivotal publications and the most prominent writers who have contributed to the development of service-oriented methods in manufacturing. The technology may include cloud-based services for data storage

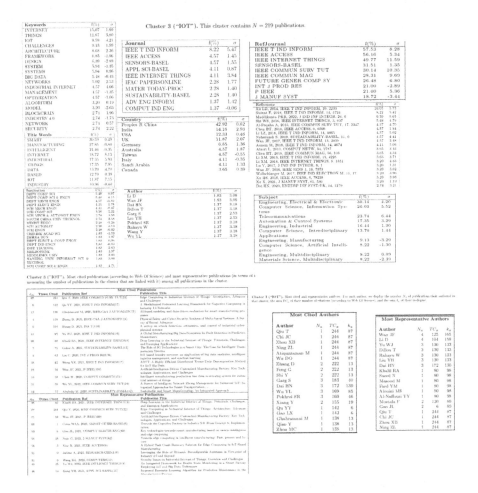

FIGURE 2.4   Cluster 3 bibliometric analysis report.

and processing, as well as Software-as-a-Service (SaaS) platforms for industrial resource planning and customer relationship management. This study might help identify the most significant services that have transformed industrial processes, as well as the authors and journals most closely linked with them. A trend toward cloud-based solutions in this cluster's literature may imply a greater emphasis on user-centric services. Furthermore, the examination will disclose the contributions made by prominent scholars to the growth of service innovation. Figure 2.5 shows the exact outcomes for Cluster 4.

2.4.2.6   Cluster 5 (Gateway)

This cluster may be smaller than most (only 57 articles), but it is very important for connecting smart manufacturing systems. We could see how gateway technologies have changed over time and how they are used in manufacturing systems by looking

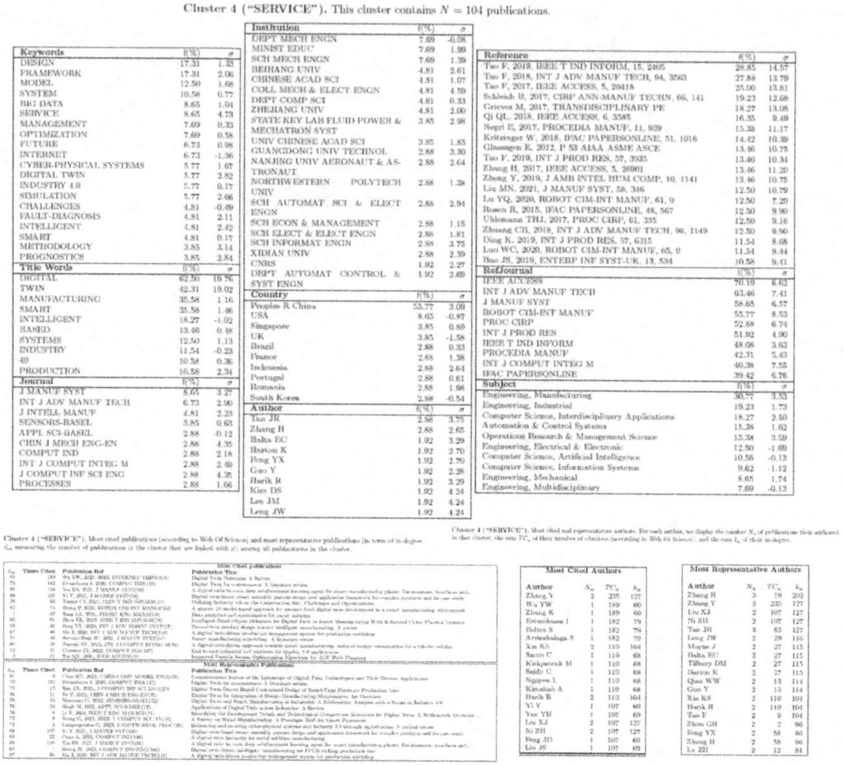

Cluster 4 ("SERVICE"). This cluster contains $N = 104$ publications.

FIGURE 2.5   Cluster 4 bibliometric analysis report.

at citations and publication years. This would help us find the most important break-throughs. These gateways make it easier to manage and move data, and they also make it possible for different technologies to work together without any problems. Some problems and ways to solve them in gateway technology are security, interoperability, and data throughput. This technology also does a lot of research on these topics. Based on citation networks, this study could show important rules or standards. As shown by publication trends, gateway technology is changing from being focused on hardware to being focused on software. This makes it easier to find leading experts and groundbreaking studies in the field. Figure 2.6 shows the specific results for Cluster 5.

2.4.2.7   Cluster 6 (Industry 4.0)

There are 61 articles in this cluster that communicate about Industry 4.0 principles. These indicators could help find important studies, the papers that have been cited the most, and the authors who have had a big impact on the history of Industry 4.0 in the manufacturing sector. Cyber-physical systems, the Internet of Things (IoT), cloud computing, cognitive computing, and other topics might be covered in articles in this group. This cluster's analysis will help us understand how Industry 4.0 is

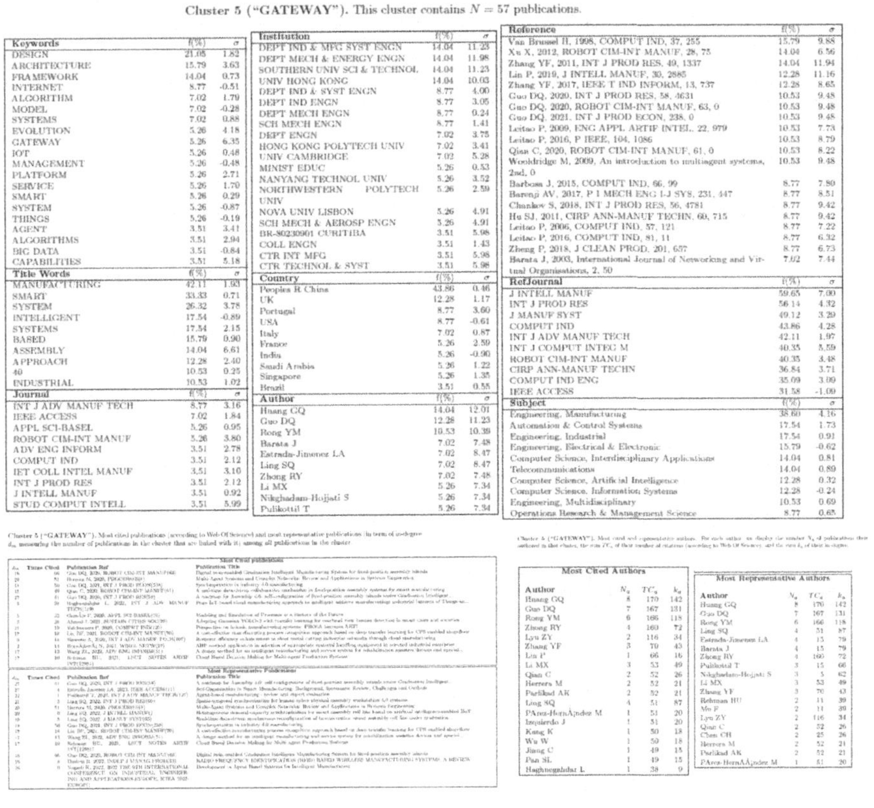
Cluster 5 ("GATEWAY"). This cluster contains $N = 57$ publications.

Keywords	f(%)	σ
DESIGN	21.05	1.82
ARCHITECTURE	15.79	3.63
FRAMEWORK	14.04	0.73
INTERNET	8.77	-0.51
ALGORITHM	7.02	1.79
MODEL	7.02	-0.28
SYSTEMS	7.02	0.88
EVOLUTION	5.26	4.18
GATEWAY	5.26	6.35
IOT	5.26	0.48
MANAGEMENT	5.26	-0.48
PLATFORM	5.26	2.71
SERVICE	5.26	1.70
SMART	5.26	0.29
SYSTEM	5.26	-0.87
THINGS	5.26	-0.19
AGENT	3.51	3.41
ALGORITHMS	3.51	2.94
BIG DATA	3.51	-0.84
CAPABILITIES	3.51	5.18

Title Words	f(%)	σ
MANUFACTURING	42.11	1.93
SMART	33.33	0.71
SYSTEM	26.32	3.78
INTELLIGENT	17.54	-0.89
SYSTEMS	17.54	2.15
BASED	15.79	0.90
ASSEMBLY	14.04	6.61
APPROACH	12.28	2.40
40	10.53	0.25
INDUSTRIAL	10.53	1.02

Journal	f(%)	σ
INT J ADV MANUF TECH	8.77	3.16
IEEE ACCESS	7.02	1.84
APPL SCI-BASEL	5.26	0.95
ROBOT CIM-INT MANUF	5.26	3.80
ADV ENG INFORM	3.51	2.78
COMPUT IND	3.51	2.12
IET COLL INTEL MANUF	3.51	3.10
INT J PROD RES	3.51	2.12
J INTELL MANUF	3.51	0.92
STUD COMPUT INTELL	3.51	5.99

Institution	f(%)	σ
DEPT IND & MFG SYST ENGN	14.04	11.23
DEPT MECH & ENERGY ENGN	14.04	11.98
SOUTHERN UNIV SCI & TECHNOL	14.04	11.23
UNIV HONG KONG	14.04	10.03
DEPT IND & SYST ENGN	8.77	4.00
DEPT IND ENGN	8.77	3.05
DEPT MECH ENGN	8.77	0.24
SCH MECH ENGN	8.77	1.41
DEPT ENGN	7.02	3.75
HONG KONG POLYTECH UNIV	7.02	3.41
UNIV CAMBRIDGE	7.02	5.28
MINIST EDUC	5.26	0.53
NANYANG TECHNOL UNIV	5.26	3.52
NORTHWESTERN POLYTECH UNIV	5.26	2.59
NOVA UNIV LISBON	5.26	4.91
SCH MECH & AEROSP ENGN	5.26	4.91
BR-80230901 CURITIBA	3.51	5.98
COLL ENGN	3.51	1.43
CTR INT MFG	3.51	5.98
CTR TECHNOL & SYST	3.51	5.98

Country	f(%)	σ
Peoples R China	43.86	0.46
UK	12.28	1.17
Portugal	8.77	3.60
USA	8.77	-0.61
Italy	7.02	0.87
France	5.26	2.59
India	5.26	-0.90
Saudi Arabia	5.26	1.22
Singapore	5.26	1.35
Brazil	3.51	0.55

Author	f(%)	σ
Huang GQ	14.04	12.01
Guo DQ	12.28	11.23
Rong YM	10.53	10.39
Barata J	7.02	7.48
Estrada-Jimenez LA	7.02	8.47
Ling SQ	7.02	8.47
Zhong RY	7.02	7.48
Li MX	5.26	7.34
Nikghadam-Hojjati S	5.26	7.34
Pulikottil T	5.26	7.34

Reference	f(%)	σ
Van Brussel H, 1998, COMPUT IND, 37, 255	15.79	9.88
Xu X, 2012, ROBOT CIM-INT MANUF, 28, 75	14.04	6.56
Zhang YF, 2011, INT J PROD RES, 49, 1337	14.04	11.94
Lin P, 2019, J INTELL MANUF, 30, 2885	12.28	11.16
Zhang YF, 2017, IEEE T IND INFORM, 13, 737	12.28	8.65
Guo DQ, 2020, INT J PROD RES, 58, 4631	10.53	9.48
Guo DQ, 2020, ROBOT CIM-INT MANUF, 63, 0	10.53	9.48
Guo DQ, 2021, INT J PROD ECON, 238, 0	10.53	9.48
Leitao P, 2009, ENG APPL ARTIF INTEL, 22, 979	10.53	7.73
Leitao P, 2016, P IEEE, 104, 1086	10.53	8.79
Qian C, 2020, ROBOT CIM-INT MANUF, 61, 0	10.53	8.22
Wooldridge M, 2009, An introduction to multiagent systems, 2nd, 0	10.53	9.48
Barbosa J, 2015, COMPUT IND, 66, 99	8.77	7.80
Barenji AV, 2017, P I MECH ENG B-J ENG, 231, 447	8.77	8.51
Chankov S, 2018, INT J PROD RES, 56, 4781	8.77	9.42
Hu SJ, 2011, CIRP ANN-MANUF TECHN, 60, 715	8.77	9.42
Leitao P, 2006, COMPUT IND, 57, 121	8.77	7.22
Leitao P, 2016, COMPUT IND, 81, 11	8.77	6.32
Zheng P, 2018, J CLEAN PROD, 201, 657	8.77	6.73
Barata J, 2003, International Journal of Networking and Virtual Organisations, 2, 50	7.02	7.44

RefJournal	f(%)	σ
J INTELL MANUF	59.65	7.00
INT J PROD RES	56.14	4.32
J MANUF SYST	49.12	3.29
COMPUT IND	43.86	4.28
INT J ADV MANUF TECH	42.11	1.97
INT J COMPUT INTEG M	40.35	5.59
ROBOT CIM-INT MANUF	40.35	3.48
CIRP ANN-MANUF TECHN	36.84	3.71
COMPUT IND ENG	35.09	3.09
IEEE ACCESS	31.58	-1.09

Subject	f(%)	σ
Engineering, Manufacturing	38.60	4.16
Automation & Control Systems	17.54	1.73
Engineering, Industrial	17.54	0.91
Engineering, Electrical & Electronic	15.79	-0.62
Computer Science, Interdisciplinary Applications	14.04	0.81
Telecommunications	14.04	0.89
Computer Science, Artificial Intelligence	12.28	0.32
Computer Science, Information Systems	12.28	-0.24
Engineering, Multidisciplinary	10.53	0.69
Operations Research & Management Science	8.77	0.65

Most Cited Authors	N_a	TC_a	k_a
Huang GQ	8	170	142
Guo DQ	7	167	131
Rong YM	6	166	118
Zhong RY	4	160	72
Lyu ZY	2	116	34
Zhang YF	3	70	43
Lin P	1	66	16
Li MX	3	53	49
Qian C	2	52	26
Herrera M	2	52	21
Parlikad AK	2	52	21
Ling SQ	4	51	87
PArez-HernĂ¡ndez M	1	51	20
Izquierdo J	1	51	20
Kong K	1	50	18
Wu W	1	50	18
Jiang C	1	49	15
Pan SL	1	49	15
Haghnegahdar L	1	38	9

Most Representative Authors	N_a	TC_a	k_a
Huang GQ	8	170	142
Guo DQ	7	167	131
Rong YM	6	166	118
Ling SQ	4	51	87
Estrada-Jimenez LA	4	15	79
Barata J	4	15	79
Zhong RY	4	160	72
Pulikottil T	3	15	66
Nikghadam-Hojjati S	3	5	62
Li MX	3	53	49
Zhang YF	3	70	43
Rehman HU	2	11	39
Mo F	2	11	39
Lyu ZY	2	116	34
Qian C	2	52	26
Chen CH	2	25	26
Herrera M	2	52	21
Parlikad AK	2	52	21
PArez-HernĂ¡ndez M	1	51	20

FIGURE 2.6 Cluster 5 bibliometric analysis report.

being used and adopted. You can easily find the paradigm's foundational research and new ideas by keeping an eye on citations and authorship patterns. In the cluster, IoT, data analytics, and cyber-physical systems would all combine. Furthermore, the bibliometric data will show how research on Industry 4.0 is spread across regions, showing where the Fourth Industrial Revolution is really happening. Figure 2.7 displays the specific results for Cluster 6.

2.4.2.8 Cluster 7 (Heuristic)

The 53 articles in this group point to a niche that is mostly about heuristic ways to make things. Along with the bibliometric information, the h-index and the average number of references could show how good and important the cluster's research was, pointing out important contributions and trends in using heuristics to solve manufacturing problems. Heuristics are great for dealing with tough issues when standard methods don't work. You could use ant colony optimisation, genetic algorithms, or simulated annealing to plan production, divide up resources, or design systems. In the literature, explicit, adaptive heuristics are replacing old ways of solving problems. Many papers that have been cited show that heuristics work well in operational

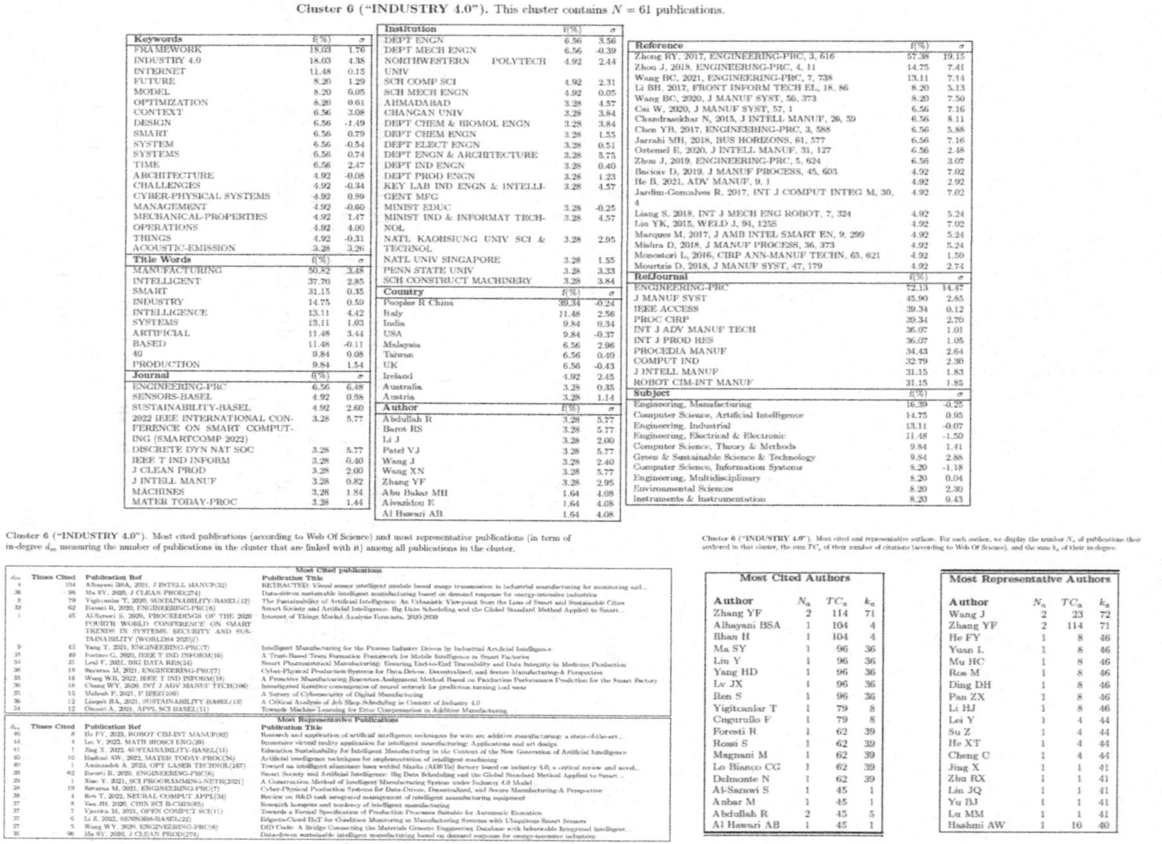

Cluster 6 ("INDUSTRY 4.0"). This cluster contains $N = 61$ publications.

Keywords	f(%)	σ
FRAMEWORK	18.03	1.76
INDUSTRY 4.0	18.03	4.38
INTERNET	11.48	0.15
FUTURE	8.20	1.29
MODEL	8.20	0.05
OPTIMIZATION	8.20	0.61
CONTEXT	6.56	3.08
DESIGN	6.56	-1.49
SMART	6.56	0.79
SYSTEM	6.56	-0.54
SYSTEMS	6.56	0.74
TIME	6.56	2.47
ARCHITECTURE	4.92	-0.08
CHALLENGES	4.92	-0.34
CYBER-PHYSICAL SYSTEMS	4.92	0.89
MANAGEMENT	4.92	-0.60
MECHANICAL-PROPERTIES	4.92	1.47
OPERATIONS	4.92	4.00
THINGS	4.92	-0.31
ACOUSTIC-EMISSION	3.28	3.26

Title Words	f(%)	σ
MANUFACTURING	50.82	3.48
INTELLIGENT	37.70	2.85
SMART	31.15	0.35
INDUSTRY	14.75	0.59
INTELLIGENCE	13.11	4.42
SYSTEMS	13.11	1.03
ARTIFICIAL	11.48	3.44
BASED	11.48	-0.11
40	9.84	0.08
PRODUCTION	9.84	1.54

Journal	f(%)	σ
ENGINEERING-PRC	6.56	6.48
SENSORS-BASEL	4.92	0.98
SUSTAINABILITY-BASEL	4.92	2.60
2022 IEEE INTERNATIONAL CONFERENCE ON SMART COMPUTING (SMARTCOMP 2022)	3.28	5.77
DISCRETE DYN NAT SOC	3.28	5.77
IEEE T IND INFORM	3.28	0.40
J CLEAN PROD	3.28	2.00
J INTELL MANUF	3.28	0.82
MACHINES	3.28	1.84
MATER TODAY-PROC	3.28	1.44

Institution	f(%)	σ
DEPT ENGN	6.56	3.56
DEPT MECH ENGN	6.56	-0.39
NORTHWESTERN POLYTECH UNIV	4.92	2.44
SCH COMP SCI	4.92	2.31
SCH MECH ENGN	4.92	0.05
AHMADABAD	3.28	4.57
CHANGAN UNIV	3.28	3.84
DEPT CHEM & BIOMOL ENGN	3.28	3.84
DEPT CHEM ENGN	3.28	1.55
DEPT ELECT ENGN	3.28	0.51
DEPT ENGN & ARCHITECTURE	3.28	5.75
DEPT IND ENGN	3.28	0.40
DEPT PROD ENGN	3.28	1.23
KEY LAB IND ENGN & INTELLIGENT MFG	3.28	4.57
MINIST EDUC	3.28	-0.25
MINIST IND & INFORMAT TECHNOL	3.28	4.57
NATL KAOHSIUNG UNIV SCI & TECHNOL	3.28	2.95
NATL UNIV SINGAPORE	3.28	1.55
PENN STATE UNIV	3.28	3.33
SCH CONSTRUCT MACHINERY	3.28	3.84

Country	f(%)	σ
Peoples R China	39.34	-0.24
Italy	11.48	2.56
India	9.84	0.34
USA	9.84	-0.37
Malaysia	6.56	2.96
Taiwan	6.56	0.49
UK	6.56	-0.43
Ireland	4.92	2.45
Australia	3.28	0.35
Austria	3.28	1.14

Author	f(%)	σ
Abdullah R	3.28	5.77
Barot RS	3.28	5.77
Li J	3.28	2.00
Patel VJ	3.28	5.77
Wang J	3.28	2.40
Wang XN	3.28	5.77
Zhang YF	3.28	2.95
Abu Bakar MH	1.64	4.08
Aivazidou E	1.64	4.08
Al Hawari AB	1.64	4.08

Reference	f(%)	σ
Zhong RY, 2017, ENGINEERING-PRC, 3, 616	57.38	19.15
Zhou J, 2018, ENGINEERING-PRC, 4, 11	14.75	7.41
Wang BC, 2021, ENGINEERING-PRC, 7, 738	13.11	7.14
Li BH, 2017, FRONT INFORM TECH EL, 18, 86	8.20	5.13
Wang BC, 2020, J MANUF SYST, 56, 373	8.20	7.50
Cai W, 2020, J MANUF SYST, 57, 1	6.56	7.16
Chandrasekhar N, 2015, J INTELL MANUF, 26, 59	6.56	8.11
Chen YB, 2017, ENGINEERING-PRC, 3, 588	6.56	5.88
Jarrahi MH, 2018, BUS HORIZONS, 61, 577	6.56	7.16
Oztemel E, 2020, J INTELL MANUF, 31, 127	6.56	2.48
Zhou J, 2019, ENGINEERING-PRC, 5, 624	6.56	3.07
Baciov D, 2019, J MANUF PROCESS, 45, 603	4.92	7.02
He B, 2021, ADV MANUF, 9, 1	4.92	2.92
Jardim-Goncalves R, 2017, INT J COMPUT INTEG M, 30, 4	4.92	7.02
Liang S, 2018, INT J MECH ENG ROBOT, 7, 324	4.92	5.24
Lin YK, 2015, WELD J, 94, 125S	4.92	7.02
Marques M, 2017, J AMB INTEL SMART EN, 9, 299	4.92	5.24
Mishra D, 2018, J MANUF PROCESS, 36, 373	4.92	5.24
Monostori L, 2016, CIRP ANN-MANUF TECHN, 65, 621	4.92	1.50
Mourtzis D, 2018, J MANUF SYST, 47, 179	4.92	2.74

RefJournal	f(%)	σ
ENGINEERING-PRC	72.13	14.47
J MANUF SYST	45.90	2.85
IEEE ACCESS	39.34	0.12
PROC CIRP	39.34	2.70
INT J ADV MANUF TECH	36.07	1.01
INT J PROD RES	36.07	1.05
PROCEDIA MANUF	34.43	2.64
COMPUT IND	32.79	2.30
J INTELL MANUF	31.15	1.83
ROBOT CIM-INT MANUF	31.15	1.85

Subject	f(%)	σ
Engineering, Manufacturing	16.39	-0.25
Computer Science, Artificial Intelligence	14.75	0.95
Engineering, Industrial	13.11	-0.07
Engineering, Electrical & Electronic	11.48	-1.50
Computer Science, Theory & Methods	9.84	1.41
Green & Sustainable Science & Technology	9.84	2.88
Computer Science, Information Systems	8.20	-1.18
Engineering, Multidisciplinary	8.20	0.04
Environmental Sciences	8.20	2.30
Instruments & Instrumentation	8.20	0.43

Cluster 6 ("INDUSTRY 4.0"). Most cited publications (according to Web Of Science) and most representative publications (in term of in-degree d_{in} measuring the number of publications in the cluster that are linked with it) among all publications in the cluster.

Most Cited publications

d_{in}	Times Cited	Publication Ref	Publication Title
4	104	Alhayani BSA, 2021, J INTELL MANUF(32)	RETRACTED: Visual sensor intelligent module based image transmission in industrial manufacturing for monitoring and...
36	96	Ma SY, 2020, J CLEAN PROD(274)	Data-driven sustainable intelligent manufacturing based on demand response for energy-intensive industries
8	79	Yigitcanlar T, 2020, SUSTAINABILITY-BASEL(12)	The Sustainability of Artificial Intelligence: An Urbanistic Viewpoint from the Lens of Smart and Sustainable Cities
39	62	Foresti R, 2020, ENGINEERING-PRC(6)	Smart Society and Artificial Intelligence: Big Data Scheduling and the Global Standard Method Applied to Smart...
1	45	Al-Sarawi S, 2020, PROCEEDINGS OF THE 2020 FOURTH WORLD CONFERENCE ON SMART TRENDS IN SYSTEMS, SECURITY AND SUSTAINABILITY (WORLDS4 2020)()	Internet of Things Market Analysis Forecasts, 2020-2030
9	43	Yang T, 2021, ENGINEERING-PRC(7)	Intelligent Manufacturing for the Process Industry Driven by Industrial Artificial Intelligence
35	40	Fortino G, 2020, IEEE T IND INFORM(16)	A Trust-Based Team Formation Framework for Mobile Intelligence in Smart Factories
34	31	Leal F, 2021, BIG DATA RES(24)	Smart Pharmaceutical Manufacturing: Ensuring End-to-End Traceability and Data Integrity in Medicine Production
38	19	Suvarna M, 2021, ENGINEERING-PRC(7)	Cyber-Physical Production Systems for Data-Driven, Decentralized, and Secure Manufacturing-A Perspective
35	18	Wang WB, 2022, IEEE T IND INFORM(18)	A Proactive Manufacturing Resources Assignment Method Based on Production Performance Prediction for the Smart Factory
36	18	Chang WY, 2020, INT J ADV MANUF TECH(106)	Investigated iterative convergences of neural network for prediction turning tool wear
35	15	Mahesh P, 2021, P IEEE(109)	A Survey of Cybersecurity of Digital Manufacturing
36	12	Lisquit RA, 2021, SUSTAINABILITY-BASEL(13)	A Critical Analysis of Job Shop Scheduling in Context of Industry 4.0
34	12	Omairi A, 2021, APPL SCI-BASEL(11)	Towards Machine Learning for Error Compensation in Additive Manufacturing

Most Representative Publications

d_{in}	Times Cited	Publication Ref	Publication Title
46	8	He FY, 2023, ROBOT CIM-INT MANUF(82)	Research and application of artificial intelligence techniques for wire arc additive manufacturing: a state-of-the-art...
44	4	Lei Y, 2023, MATH BIOSCI ENG(20)	Immersive virtual reality application for intelligent manufacturing: Applications and art design
41	1	Jing X, 2022, SUSTAINABILITY-BASEL(14)	Education Sustainability for Intelligent Manufacturing in the Context of the New Generation of Artificial Intelligence
40	10	Hashmi AW, 2022, MATER TODAY-PROC(56)	Artificial intelligence techniques for implementation of intelligent machining
40	1	Aminzadeh A, 2023, OPT LASER TECHNOL(167)	Toward an intelligent aluminum laser welded blanks (ALWBs) factory based on industry 4.0; a critical review and novel...
39	62	Foresti R, 2020, ENGINEERING-PRC(6)	Smart Society and Artificial Intelligence: Big Data Scheduling and the Global Standard Method Applied to Smart...
39	1	Xiao Y, 2021, SCI PROGRAMMING-NETH(2021)	A Construction Method of Intelligent Manufacturing System under Industry 4.0 Model
38	19	Suvarna M, 2021, ENGINEERING-PRC(7)	Cyber-Physical Production Systems for Data-Driven, Decentralized, and Secure Manufacturing-A Perspective
38	4	Ren T, 2022, NEURAL COMPUT APPL(34)	Review on R&D task integrated management of intelligent manufacturing equipment
37	8	Yan JH, 2020, CHIN SCI B-CHIN(65)	Research hotspots and tendency of intelligent manufacturing
37	7	Vjestica M, 2021, OPEN COMPUT SCI(11)	Towards a Formal Specification of Production Processes Suitable for Automatic Execution
37	6	Li Z, 2022, SENSORS-BASEL(22)	Edge-to-Cloud IIoT for Condition Monitoring in Manufacturing Systems with Ubiquitous Smart Sensors
37	5	Wang WY, 2020, ENGINEERING-PRC(6)	DID Code: A Bridge Connecting the Materials Genome Engineering Database with Inheritable Integrated Intelligent...
36	96	Ma SY, 2020, J CLEAN PROD(274)	Data-driven sustainable intelligent manufacturing based on demand response for energy-intensive industries

Cluster 6 ("INDUSTRY 4.0"). Most cited and representative authors. For each author, we display the number N_a of publications they authored in that cluster, the sum TC_a of their number of citations (according to Web Of Science), and the sum k_a of their in-degree.

Most Cited Authors

Author	N_a	TC_a	k_a
Zhang YF	2	114	71
Alhayani BSA	1	104	4
Ilhan H	1	104	4
Ma SY	1	96	36
Liu Y	1	96	36
Yang HD	1	96	36
Lv JX	1	96	36
Ren S	1	96	36
Yigitcanlar T	1	79	8
Cugurullo F	1	79	8
Foresti R	1	62	39
Rossi S	1	62	39
Magnani M	1	62	39
Lo Bianco CG	1	62	39
Delmonte N	1	62	39
Al-Sarawi S	1	45	1
Anbar M	1	45	1
Abdullah R	2	45	5
Al Hawari AB	1	45	1

Most Representative Authors

Author	N_a	TC_a	k_a
Wang J	2	23	72
Zhang YF	2	114	71
He FY	1	8	46
Yuan L	1	8	46
Mu HC	1	8	46
Ros M	1	8	46
Ding DH	1	8	46
Pan ZX	1	8	46
Li HJ	1	8	46
Lei Y	1	4	44
Su Z	1	4	44
He XT	1	4	44
Cheng C	1	4	44
Jing X	1	1	41
Zhu RX	1	1	41
Lin JQ	1	1	41
Yu BJ	1	1	41
Lu MM	1	1	41
Hashmi AW	1	10	40

FIGURE 2.7 Cluster 6 bibliometric analysis report.

optimisation by using new methods or benchmark case studies. Figure 2.8 shows the specific results for Cluster 7.

2.4.2.9 Cluster 8 (Cutting Force)

Among the smaller clusters, this one contains 25 articles, which indicates a highly specialised area of study within manufacturing. By analysing the metrics, we could see which research areas are most important, which papers are most influential, and who the top authors are. We could see how cutting forces play a role in understanding and optimising manufacturing processes. Measurement, modelling, and controlling cutting forces are likely to dominate this cluster. This study will help identify the most impactful research, characterised by high citation counts that have significantly advanced cutting force understanding or application. Figure 2.9 illustrates the specific outcomes for Cluster 8.

2.4.2.10 Cluster 9 (Remote Maintenance)

There are 14 articles in this cluster that may cover emerging trends and technologies in remote maintenance. A key factor in manufacturing is maintenance, which impacts productivity and uptime. The strategies and technologies for remote maintenance are likely in Cluster 9. Predictive maintenance, where data analytics can predict equipment failures before they happen, and condition monitoring systems that let maintenance teams monitor equipment status remotely, would be especially relevant. Dissecting the research landscape will reveal the most critical remote diagnostics and maintenance technologies. Figure 2.10 illustrates the specific outcomes for Cluster 9.

2.4.2.11 Cluster 10 (Fabrication)

With 175 articles, the fabrication cluster indicates that smart manufacturing is a significant area of focus. Through the use of bibliometric analysis, we can gain insight into the key technologies, methods, and materials shaping modern manufacturing practices. This research's publications and other metrics could reveal how current and impactful it is, highlighting significant contributions and trends. These technologies are likely to revolutionise product design, customisation, and production as a result of this analysis. Moreover, it may include a listing of the most cited authors and publications, as well as highlighting emerging trends and future directions in the field. It would highlight key innovations, benchmark studies, and leading voices in advanced fabrication by finding the most cited and representative publications. Figure 2.11 illustrates the specific outcomes for Cluster 10.

Based on the bibliometric data from an image, the detailed analysis of each cluster would illustrate key research areas, influential research, and emerging trends in intelligent and smart manufacturing.

2.4.2.12 Thematic Analysis

Using thematic analysis, you can see how diverse each cluster is. For instance, Cluster 1 (CNN) explores convolutional neural networks in manufacturing, and Cluster 5 (Gateway) explores communication and system integration in smart manufacturing.

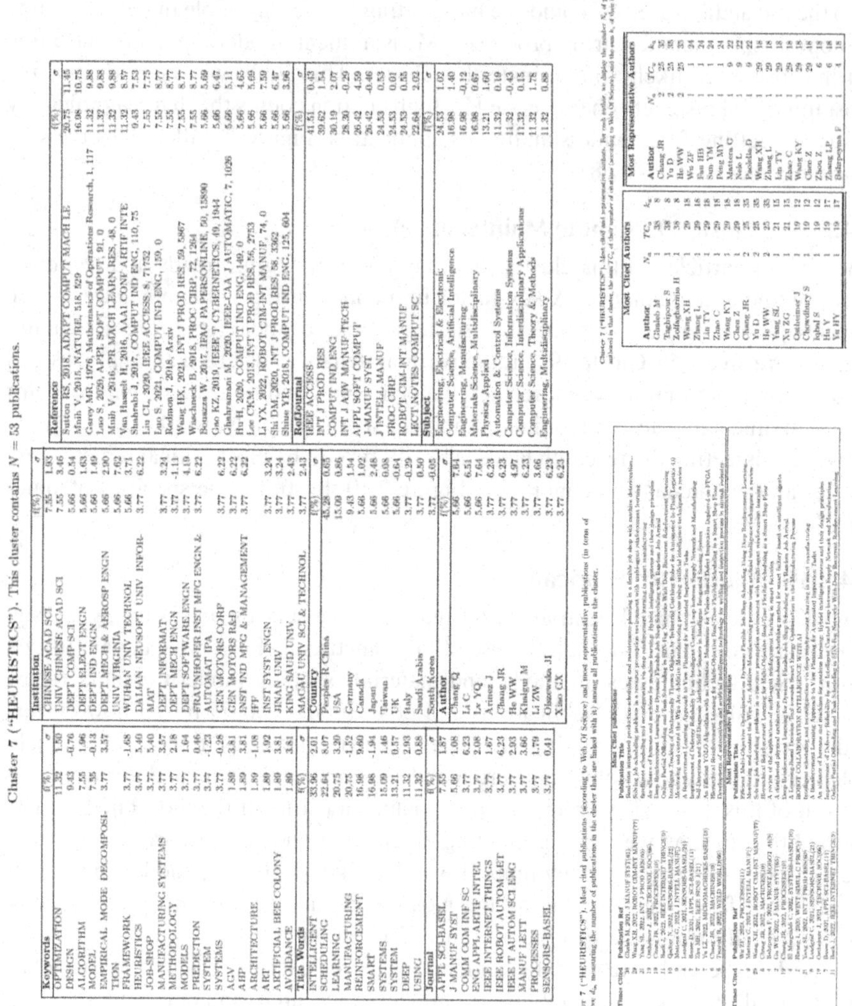

FIGURE 2.8 Cluster 7 bibliometric analysis report.

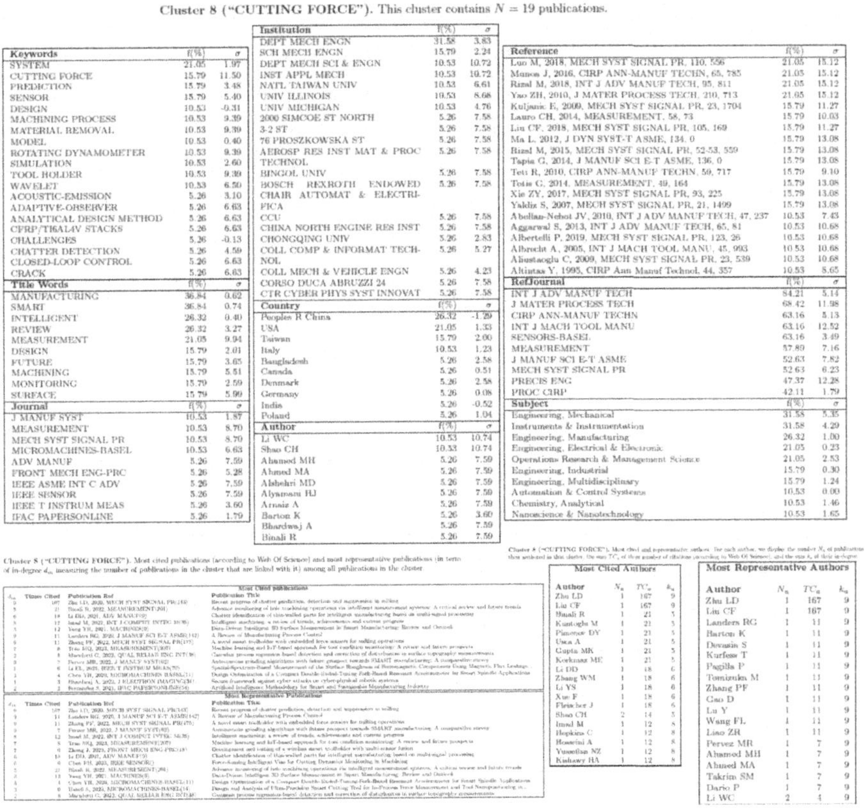

Cluster 8 ("CUTTING FORCE"). This cluster contains $N = 19$ publications.

Keywords	f(%)	σ
SYSTEM	21.05	1.97
CUTTING FORCE	15.79	11.50
PREDICTION	15.79	3.48
SENSOR	15.79	5.40
DESIGN	10.53	-0.31
MACHINING PROCESS	10.53	9.39
MATERIAL REMOVAL	10.53	9.39
MODEL	10.53	0.40
ROTATING DYNAMOMETER	10.53	9.39
SIMULATION	10.53	2.60
TOOL HOLDER	10.53	9.39
WAVELET	10.53	6.50
ACOUSTIC-EMISSION	5.26	3.10
ADAPTIVE-OBSERVER	5.26	6.63
ANALYTICAL DESIGN METHOD	5.26	6.63
CFRP/TI6AL4V STACKS	5.26	6.63
CHALLENGES	5.26	-0.13
CHATTER DETECTION	5.26	4.59
CLOSED-LOOP CONTROL	5.26	6.63
CRACK	5.26	6.63
Title Words	f(%)	σ
MANUFACTURING	36.84	0.62
SMART	36.84	0.74
INTELLIGENT	26.32	0.40
REVIEW	26.32	3.27
MEASUREMENT	21.05	9.94
DESIGN	15.79	2.01
FUTURE	15.79	3.65
MACHINING	15.79	5.51
MONITORING	15.79	2.59
SURFACE	15.79	5.99
Journal	f(%)	σ
J MANUF SYST	10.53	1.87
MEASUREMENT	10.53	8.70
MECH SYST SIGNAL PR	10.53	8.70
MICROMACHINES-BASEL	10.53	6.63
ADV MANUF	5.26	7.59
FRONT MECH ENG-PRC	5.26	5.28
IEEE ASME INT C ADV	5.26	7.59
IEEE SENSOR	5.26	7.59
IEEE T INSTRUM MEAS	5.26	3.60
IFAC PAPERSONLINE	5.26	1.79

Institution	f(%)	σ
DEPT MECH ENGN	31.58	3.83
SCH MECH ENGN	15.79	2.24
DEPT MECH SCI & ENGN	10.53	10.72
INST APPL MECH	10.53	10.72
NATL TAIWAN UNIV	10.53	6.61
UNIV ILLINOIS	10.53	8.68
UNIV MICHIGAN	10.53	4.76
2000 SIMCOE ST NORTH	5.26	7.58
3-2 ST	5.26	7.58
76 PROSZKOWSKA ST	5.26	7.58
AEROSP RES INST MAT & PROC TECHNOL	5.26	7.58
BINGOL UNIV	5.26	7.58
BOSCH REXROTH ENDOWED CHAIR AUTOMAT & ELECTRIFICA	5.26	7.58
CCU	5.26	7.58
CHINA NORTH ENGINE RES INST	5.26	7.58
CHONGQING UNIV	5.26	2.83
COLL COMP & INFORMAT TECHNOL	5.26	5.27
COLL MECH & VEHICLE ENGN	5.26	4.23
CORSO DUCA ABRUZZI 24	5.26	7.58
CTR CYBER PHYS SYST INNOVAT	5.26	7.58
Country	f(%)	σ
Peoples R China	26.32	-1.29
USA	21.05	1.33
Taiwan	15.79	2.00
Italy	10.53	1.23
Bangladesh	5.26	2.58
Canada	5.26	0.51
Denmark	5.26	2.58
Germany	5.26	0.08
India	5.26	-0.52
Poland	5.26	1.04
Author	f(%)	σ
Li WC	10.53	10.74
Shao CH	10.53	10.74
Ahamed MH	5.26	7.59
Ahmed MA	5.26	7.59
Alshehri MD	5.26	7.59
Alyamani HJ	5.26	7.59
Arnaiz A	5.26	7.59
Barton K	5.26	3.60
Bhardwaj A	5.26	7.59
Binali R	5.26	7.59

Reference	f(%)	σ
Luo M, 2018, MECH SYST SIGNAL PR, 110, 556	21.05	15.12
Manoa J, 2016, CIRP ANN-MANUF TECHN, 65, 785	21.05	15.12
Rizal M, 2018, INT J ADV MANUF TECH, 95, 811	21.05	15.12
Yao ZH, 2010, J MATER PROCESS TECH, 210, 713	21.05	15.12
Kuljanic E, 2009, MECH SYST SIGNAL PR, 23, 1704	15.79	11.27
Lauro CH, 2014, MEASUREMENT, 58, 73	15.79	10.03
Liu CF, 2018, MECH SYST SIGNAL PR, 105, 169	15.79	11.27
Ma L, 2012, J DYN SYST-T ASME, 134, 0	15.79	13.08
Rizal M, 2015, MECH SYST SIGNAL PR, 52-53, 559	15.79	13.08
Tapia G, 2014, J MANUF SCI E-T ASME, 136, 0	15.79	13.08
Teti R, 2010, CIRP ANN-MANUF TECHN, 59, 717	15.79	9.10
Totis G, 2014, MEASUREMENT, 49, 164	15.79	13.08
Xie ZY, 2017, MECH SYST SIGNAL PR, 93, 225	15.79	13.08
Yaldiz S, 2007, MECH SYST SIGNAL PR, 21, 1499	15.79	13.08
Abellan-Nebot JV, 2010, INT J ADV MANUF TECH, 47, 237	10.53	7.43
Aggarwal S, 2013, INT J ADV MANUF TECH, 65, 81	10.53	10.68
Albertelli P, 2019, MECH SYST SIGNAL PR, 123, 26	10.53	10.68
Albrecht A, 2005, INT J MACH TOOL MANU, 45, 993	10.53	10.68
Altintas Y, 1995, CIRP Ann Manuf Technol, 44, 357	10.53	8.65
Aliustaoglu C, 2009, MECH SYST SIGNAL PR, 23, 539	10.53	10.68
RefJournal	f(%)	σ
INT J ADV MANUF TECH	84.21	5.14
J MATER PROCESS TECH	68.42	11.98
CIRP ANN-MANUF TECHN	63.16	5.13
INT J MACH TOOL MANU	63.16	12.52
SENSORS-BASEL	63.16	3.49
MEASUREMENT	57.89	7.16
J MANUF SCI E-T ASME	52.63	7.82
MECH SYST SIGNAL PR	52.63	6.23
PRECIS ENG	47.37	12.28
PROC CIRP	42.11	1.79
Subject	f(%)	σ
Engineering, Mechanical	31.58	5.35
Instruments & Instrumentation	31.58	4.29
Engineering, Manufacturing	26.32	1.00
Engineering, Electrical & Electronic	21.05	0.23
Operations Research & Management Science	21.05	2.53
Engineering, Industrial	15.79	0.30
Engineering, Multidisciplinary	15.79	1.24
Automation & Control Systems	10.53	0.00
Chemistry, Analytical	10.53	1.46
Nanoscience & Nanotechnology	10.53	1.65

Cluster 8 ("CUTTING FORCE"). Most cited publications (according to Web Of Science) and most representative publications (in terms of in-degree d_{in} measuring the number of publications in the cluster that are linked with it) among all publications in the cluster.

Cluster 8 ("CUTTING FORCE"). Most cited and representative authors. For each author, we display the number N_a of publications then authored in this cluster, the sum TC_a of the number of citations (according to Web Of Science), and the sum k_a of their in-degree.

Most Cited Authors		
Author	N_a	TC_a	k_a

Author	N_a	TC_a	k_a
Zhu LD	1	167	9
Liu CF	1	167	9
Binali R	1	21	5
Kuntoglu M	1	21	5
Pimenov DY	1	21	5
Usca A	1	21	5
Gupta MK	1	21	5
Korkmaz ME	1	21	5
Li DD	1	18	6
Zhang WM	1	18	6
Li YS	1	18	6
Xue F	1	18	6
Fleischer J	1	18	6
Shao CH	2	14	5
Imad M	1	12	8
Hopkins C	1	12	8
Hosseini A	1	12	8
Yussefian NZ	1	12	8
Kishawy HA	1	12	8

Most Representative Authors			
Author	N_a	TC_a	k_a
Zhu LD	1	167	9
Liu CF	1	167	9
Landers RG	1	11	9
Barton K	1	11	9
Devasia S	1	11	9
Kurfess T	1	11	9
Pagilla P	1	11	9
Tomizuka M	1	11	9
Zhang PF	1	11	9
Gao D	1	11	9
Lu Y	1	11	9
Wang FL	1	11	9
Liao ZR	1	11	9
Perves MR	1	7	9
Ahamed MH	1	7	9
Ahmed MA	1	7	9
Takrim SM	1	7	9
Dario P	1	7	9
Li WC	2	4	9

FIGURE 2.9 Cluster 8 bibliometric analysis report.

Multifaceted research in intelligent and smart manufacturing is reflected in such thematic diversity.

2.4.2.13 Evolutionary Trends

Trends and shifts in research focus over time can be seen in the temporal evolution of clusters. Apparently, Cluster 6 (Industry 4.0) has seen a significant increase in publications in recent years, reflecting the growing interest in Industry 4.0. On the other hand, Cluster 10 (Fabrication) has a consistent publication trend, indicating continued fabrication research.

2.4.3 Network and Collaboration Analysis

2.4.3.1 Authorship Networks

Collaboration patterns are revealed in authorship networks within and across clusters. Among the authors in Cluster 2 (Technologies), there's a high degree of collaboration among different institutions and countries, pointing to a global research

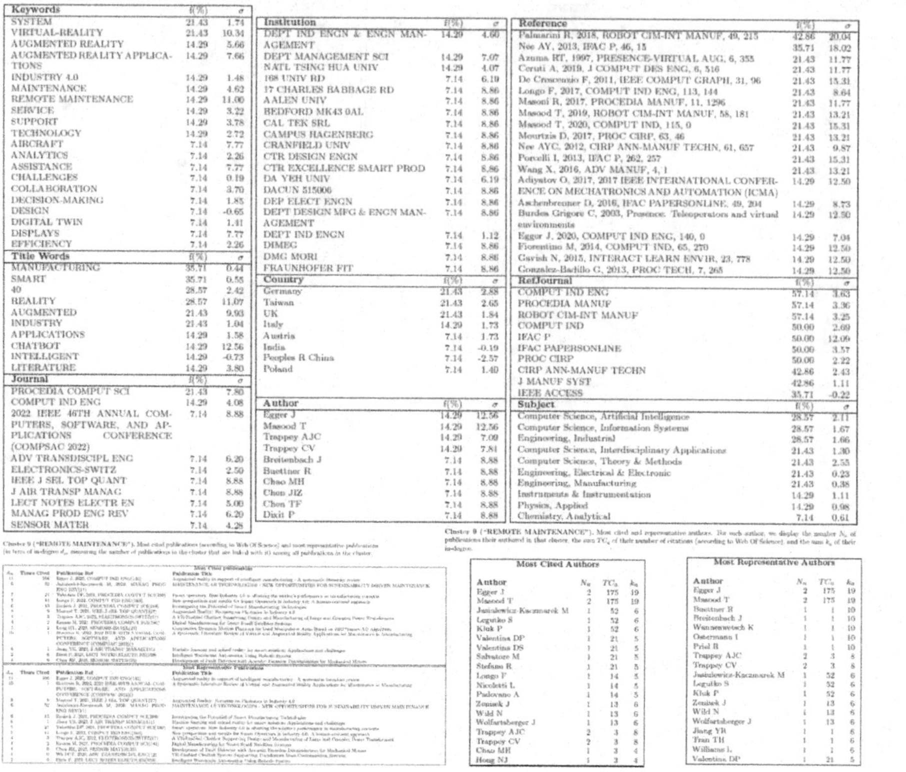

Cluster 9 ("REMOTE MAINTENANCE"). This cluster contains $N = 14$ publications.

Keywords	f(%)	σ
SYSTEM	21.43	1.74
VIRTUAL-REALITY	21.43	10.34
AUGMENTED REALITY	14.29	5.66
AUGMENTED REALITY APPLICATIONS	14.29	7.66
INDUSTRY 4.0	14.29	1.48
MAINTENANCE	14.29	4.62
REMOTE MAINTENANCE	14.29	11.00
SERVICE	14.29	3.22
SUPPORT	14.29	3.78
TECHNOLOGY	14.29	2.72
AIRCRAFT	7.14	7.77
ANALYTICS	7.14	2.26
ASSISTANCE	7.14	7.77
CHALLENGES	7.14	0.19
COLLABORATION	7.14	3.70
DECISION-MAKING	7.14	1.85
DESIGN	7.14	-0.65
DIGITAL TWIN	7.14	1.41
DISPLAYS	7.14	7.77
EFFICIENCY	7.14	2.26

Title Words	f(%)	σ
MANUFACTURING	35.71	0.44
SMART	35.71	0.55
40	28.57	2.42
REALITY	28.57	11.07
AUGMENTED	21.43	9.93
INDUSTRY	21.43	1.04
APPLICATIONS	14.29	1.58
CHATBOT	14.29	12.56
INTELLIGENT	14.29	-0.73
LITERATURE	14.29	3.80

Journal	f(%)	σ
PROCEDIA COMPUT SCI	21.43	7.80
COMPUT IND ENG	14.29	4.08
2022 IEEE 46TH ANNUAL COMPUTERS, SOFTWARE, AND APPLICATIONS CONFERENCE (COMPSAC 2022)	7.14	8.88
ADV TRANSDISCIPL ENG	7.14	6.20
ELECTRONICS-SWITZ	7.14	2.50
IEEE J SEL TOP QUANT	7.14	8.88
J AIR TRANSP MANAG	7.14	8.88
LECT NOTES ELECTR EN	7.14	5.00
MANAG PROD ENG REV	7.14	6.20
SENSOR MATER	7.14	4.28

Institution	f(%)	σ
DEPT IND ENGN & ENGN MANAGEMENT	14.29	4.60
DEPT MANAGEMENT SCI	14.29	7.07
NATL TSING HUA UNIV	14.29	4.07
168 UNIV RD	7.14	6.19
17 CHARLES BABBAGE RD	7.14	8.86
AALEN UNIV	7.14	8.86
BEDFORD MK43 0AL	7.14	8.86
CAL TEK SRL	7.14	8.86
CAMPUS HAGENBERG	7.14	8.86
CRANFIELD UNIV	7.14	8.86
CTR DESIGN ENGN	7.14	8.86
CTR EXCELLENCE SMART PROD	7.14	8.86
DA YEH UNIV	7.14	6.19
DACUN 515006	7.14	8.86
DEP ELECT ENGN	7.14	8.86
DEPT DESIGN MFG & ENGN MANAGEMENT	7.14	8.86
DEPT IND ENGN	7.14	1.12
DIMEG	7.14	8.86
DMG MORI	7.14	8.86
FRAUNHOFER FIT	7.14	8.86

Country	f(%)	σ
Germany	21.43	2.88
Taiwan	21.43	2.65
UK	21.43	1.84
Italy	14.29	1.73
Austria	7.14	1.73
India	7.14	-0.19
Peoples R China	7.14	-2.57
Poland	7.14	1.40

Author	f(%)	σ
Egger J	14.29	12.56
Masood T	14.29	12.56
Trappey AJC	14.29	7.09
Trappey CV	14.29	7.81
Breitenbach J	7.14	8.88
Buettner R	7.14	8.88
Chao MH	7.14	8.88
Chen JIZ	7.14	8.88
Chen TF	7.14	8.88
Dixit P	7.14	8.88

Reference	f(%)	σ
Palmarini R, 2018, ROBOT CIM-INT MANUF, 49, 215	42.86	20.04
Nee AY, 2013, IFAC P, 46, 15	35.71	18.02
Azuma RT, 1997, PRESENCE-VIRTUAL AUG, 6, 355	21.43	11.77
Ceruti A, 2019, J COMPUT DES ENG, 6, 516	21.43	11.77
De Crescenzio F, 2011, IEEE COMPUT GRAPH, 31, 96	21.43	15.31
Longo F, 2017, COMPUT IND ENG, 113, 144	21.43	8.64
Masoni R, 2017, PROCEDIA MANUF, 11, 1296	21.43	11.77
Masood T, 2019, ROBOT CIM-INT MANUF, 58, 181	21.43	13.21
Masood T, 2020, COMPUT IND, 115, 0	21.43	15.31
Mourtzis D, 2017, PROC CIRP, 63, 46	21.43	13.21
Nee AYC, 2012, CIRP ANN-MANUF TECHN, 61, 657	21.43	9.87
Porcelli I, 2013, IFAC P, 262, 257	21.43	15.31
Wang X, 2016, ADV MANUF, 4, 1	21.43	13.21
Adiyatov O, 2017, 2017 IEEE INTERNATIONAL CONFERENCE ON MECHATRONICS AND AUTOMATION (ICMA)	14.29	12.50
Aschenbrenner D, 2016, IFAC PAPERSONLINE, 49, 204	14.29	8.73
Burdea Grigore C, 2003, Presence: Teleoperators and virtual environments	14.29	12.50
Egger J, 2020, COMPUT IND ENG, 140, 0	14.29	7.04
Fiorentino M, 2014, COMPUT IND, 65, 270	14.29	12.50
Gavish N, 2015, INTERACT LEARN ENVIR, 23, 778	14.29	12.50
Gonzalez-Badillo G, 2013, PROC TECH, 7, 265	14.29	12.50

RefJournal	f(%)	σ
COMPUT IND ENG	57.14	3.63
PROCEDIA MANUF	57.14	3.36
ROBOT CIM-INT MANUF	57.14	3.25
COMPUT IND	50.00	2.69
IFAC P	50.00	12.09
IFAC PAPERSONLINE	50.00	3.57
PROC CIRP	50.00	2.22
CIRP ANN-MANUF TECHN	42.86	2.43
J MANUF SYST	42.86	1.11
IEEE ACCESS	35.71	-0.22

Subject	f(%)	σ
Computer Science, Artificial Intelligence	28.57	2.11
Computer Science, Information Systems	28.57	1.67
Engineering, Industrial	28.57	1.66
Computer Science, Interdisciplinary Applications	21.43	1.30
Computer Science, Theory & Methods	21.43	2.55
Engineering, Electrical & Electronic	21.43	0.23
Engineering, Manufacturing	21.43	0.38
Instruments & Instrumentation	14.29	1.11
Physics, Applied	14.29	0.98
Chemistry, Analytical	7.14	0.61

Cluster 9 ("REMOTE MAINTENANCE"). Most cited publications (according to Web Of Science) and most representative publications (in terms of in-degree d_a, measuring the number of publications in the cluster that are linked with it) among all publications in the cluster.

Cluster 9 ("REMOTE MAINTENANCE"). Most cited and representative authors. For each author, we display the number N_a of publications their authored in that cluster, the sum TC_a of their number of citations (according to Web Of Science), and the sum k_a of their in-degree.

Most Cited Authors

Author	N_a	TC_a	k_a
Egger J	2	175	19
Masood T	2	175	19
Jasiulewicz-Kaczmarek M	1	52	6
Legutko S	1	52	6
Kluk P	1	52	6
Valentina DP	1	21	5
Valentina DS	1	21	5
Salvatore M	1	21	5
Stefano R	1	21	5
Longo F	1	14	5
Nicoletti L	1	14	5
Padovano A	1	14	5
Zoniak J	1	13	6
Wild N	1	13	6
Wolfartsberger J	1	13	6
Trappey AJC	2	3	8
Trappey CV	2	3	8
Chao MH	1	3	4
Hong NJ	1	3	4

Most Representative Authors

Author	N_a	TC_a	k_a
Egger J	2	175	19
Masood T	2	175	19
Buettner R	1	1	10
Breitenbach J	1	1	10
Wannenwetsch K	1	1	10
Ostermann I	1	1	10
Priel R	1	1	10
Trappey AJC	2	3	8
Trappey CV	2	3	8
Jasiulewicz-Kaczmarek M	1	52	6
Legutko S	1	52	6
Kluk P	1	52	6
Zoniak J	1	13	6
Wild N	1	13	6
Wolfartsberger J	1	13	6
Jiang YR	1	1	6
Tran TH	1	1	6
Williams L	1	1	6
Valentina DP	1	21	5

FIGURE 2.10 Cluster 9 bibliometric analysis report.

network. Contrary to Cluster 7 (Heuristic), which shows a localised collaboration pattern where authors are mostly affiliated with a few key institutions.

2.4.3.2 Geographic Distribution

Different countries contribute to each cluster based on their geographical location. Cluster 4 (Services/Applications) is dominated by publications from the United States and China, demonstrating the worldwide scope of service-oriented manufacturing. Cluster 8 (Cutting Force) receives contributions from a wide range of nations in Europe, Asia, and North America.

2.4.4 Impact and Influence Analysis

2.4.4.1 Citation Metrics

A citation measure reveals the effect and influence of each cluster within the scientific community. Researchers with higher h-index values and citations per publication are more likely to be recognised for their research results and disseminated to the scientific community. Innovative system in automated manufacturing inventions

Cluster 10 ("FABRICATION"). This cluster contains $N = 175$ publications.

Keywords	f(%)	σ
DESIGN	11.43	-0.60
FABRICATION	11.43	7.26
COMPOSITES	9.71	7.12
PERFORMANCE	8.57	2.74
FILMS	6.86	6.23
MECHANICAL-PROPERTIES	5.14	2.69
3D	4.57	5.07
HYDROGEL	4.57	4.64
SHAPE-MEMORY POLYMER	4.57	5.07
TEMPERATURE	4.00	3.55
DRIVEN	3.43	2.84
HYDROGELS	3.43	4.39
NANOCOMPOSITES	3.43	4.39
TRIBOELECTRIC NANOGENERATOR	3.43	4.39
BEHAVIOR	2.86	1.49
FILM	2.86	4.00
LIGHT	2.86	4.00
POLYMER	2.86	3.47
POLYMERS	2.86	3.47
POLYURETHANE	2.86	4.00

Title Words	f(%)	σ
SMART	26.29	-0.82
INTELLIGENT	16.00	-2.05
PRINTING	15.43	11.30
4D	12.57	10.18
3D	12.00	8.33
MANUFACTURING	12.00	-5.27
BASED	10.29	-0.67
MATERIALS	10.29	6.84
APPLICATIONS	8.57	2.14
PRINTED	8.00	7.72

Journal	f(%)	σ
ACS APPL MATER INTER	2.86	3.78
ADV MATER TECHNOL-US	2.86	4.82
NANO ENERGY	2.86	4.82
CHEM ENG J	2.29	4.31
MATERIALS	2.29	3.68
SMART MATER STRUCT	2.29	4.31
ADV FUNCT MATER	1.71	3.73
ADV INTELL SYST-GER	1.71	3.73
CHIN J LASERS	1.71	3.73
CURR OPIN BIOMED ENG	1.71	3.73

Institution	f(%)	σ
MINIST EDUC	10.86	4.75
DEPT MECH ENGN	10.29	1.17
CHINESE ACAD SCI	9.71	5.18
SCH MAT SCI & ENGN	5.14	4.40
SCH MECH ENGN	4.57	-0.14
SCH ENGN	4.00	1.49
TSINGHUA UNIV	4.00	2.70
DEPT CHEM	3.43	4.73
DEPT CHEM ENGN	3.43	2.82
UNIV CHINESE ACAD SCI	3.43	1.93
BEIJING INST NANOENERGY & NANOSYST	2.86	4.80
BEIJING INST TECHNOL	2.86	2.31
DEPT BIOMED ENGN	2.86	3.38
FAC ENGN	2.86	0.67
HARBIN INST TECHNOL	2.86	3.38
HUAZHONG UNIV SCI & TECHNOL	2.86	1.61
JIANGNAN UNIV	2.86	3.38
NATL UNIV SINGAPORE	2.86	2.11
BEIHANG UNIV	2.29	0.73
CAS KEY LAB MECH BEHAV & DESIGN MAT	2.29	4.29

Country	f(%)	σ
Peoples R China	54.86	3.76
USA	13.71	0.99
India	7.43	-0.55
UK	6.86	-0.59
Singapore	4.57	1.78
Germany	4.00	-0.54
Italy	4.00	-0.38
Canada	3.43	0.18
Iran	3.43	2.25
South Korea	3.43	-0.32

Author	f(%)	σ
Leng JS	2.29	4.31
Li Y	2.29	2.49
Liu LW	2.29	4.31
Zhang L	2.29	2.22
Chen CY	1.71	3.73
Gong XL	1.71	3.73
Huang YL	1.71	3.73
Liu XH	1.71	3.73
Liu YJ	1.71	3.03
Ramakrishna S	1.71	3.03

Reference	f(%)	σ
Gladman AS, 2016, NAT MATER, 15, 413	10.86	9.38
Tibbits S, 2014, ARCHIT DESIGN, 84, 116	8.00	8.03
Momeni F, 2017, MATER DESIGN, 122, 42	7.43	7.35
Fan FR, 2012, NANO ENERGY, 1, 328	6.86	7.43
Ge Q, 2014, SMART MATER STRUCT, 23, 0	6.86	7.43
Kuang X, 2019, ADV FUNCT MATER, 29, 0	6.86	7.43
Kuang X, 2018, ACS APPL MATER INTER, 10, 7381	5.71	6.78
Zhang FH, 2019, COMPOS PART A-APPL S, 125, 0	5.71	6.78
Zhang ZZ, 2019, INT J SMART NANO MAT, 10, 205	5.71	6.78
Bakarich SE, 2015, MACROMOL RAPID COMM, 36, 1211	5.14	6.43
Choi J, 2015, 3D PRINT ADDIT MANUF, 2, 159	5.14	6.43
Ge Q, 2016, SCI REP-UK, 6, 0	5.14	5.97
Khoo ZX, 2015, VIRTUAL PHYS PROTOTY, 10, 103	5.14	6.43
Kim Y, 2018, NATURE, 558, 274	5.14	6.43
Mao YQ, 2016, SCI REP-UK, 6, 0	5.14	6.43
Ahmed A, 2021, POLYMER, 228, 0	4.57	6.06
Ding Z, 2017, SCI ADV, 3, 0	4.57	6.06
Falahati M, 2020, MATER TODAY, 40, 215	4.57	6.06
Han D, 2018, ACS APPL MATER INTER, 10, 17512	4.57	6.06
Leng JS, 2011, PROG MATER SCI, 56, 1077	4.57	6.06

RefJournal	f(%)	σ
ADV MATER	70.86	23.33
ADV FUNCT MATER	68.00	23.05
ACS APPL MATER INTER	62.29	21.58
NAT COMMUN	51.43	18.83
ACS NANO	44.00	18.18
SCI REP-UK	41.71	14.56
ADV MATER TECHNOL-US	37.14	17.54
NATURE	35.43	8.73
SENSOR ACTUAT A-PHYS	35.43	13.90
SCI ADV	34.86	16.26

Subject	f(%)	σ
Materials Science, Multidisciplinary	40.57	10.27
Nanoscience & Nanotechnology	16.57	9.34
Physics, Applied	14.86	3.75
Chemistry, Physical	14.29	10.05
Instruments & Instrumentation	10.29	1.83
Chemistry, Multidisciplinary	9.71	2.90
Polymer Science	8.00	8.09
Chemistry, Analytical	6.29	1.56
Engineering, Electrical & Electronic	5.71	-4.48
Engineering, Manufacturing	5.14	-4.33

Cluster 10 ("FABRICATION"). Most cited publications (according to Web Of Science) and most representative publications (in term of in-degree d_{in} measuring the number of publications in the cluster that are linked with it) among all publications in the cluster.

Most Cited publications

d_{in}	Times Cited	Publication Ref	Publication Title
3	336	Zeng SN, 2021, SCIENCE(373)	Hierarchical-morphology metafabric for scalable passive daytime radiative cooling
9	180	Dai J, 2020, CHEM SOC REV(49)	Printed gas sensors
61	113	Falahati M, 2020, MATER TODAY(40)	Smart polymers and nanocomposites for 3D and 4D printing
7	101	Drago E, 2020, FOODS(9)	Innovations in Smart Packaging Concepts for Food: An Extensive Review
25	78	Ntagios M, 2020, ADV INTELL SYST-GER(2)	Robotic Hands with Intrinsic Tactile Sensing via 3D Printed Soft Pressure Sensors
35	73	Jiao DJ, 2022, ACCOUNTS CHEM RES(55)	Programmable Morphing Hydrogels for Soft Actuators and Robots: From Structure Designs to Active Functions
28	70	Tao R, 2020, COMPOS STRUCT(252)	4D printed multi-stable metamaterials with mechanically tunable performance
18	59	Huang JR, 2020, NANO ENERGY(68)	A universal and arbitrary tactile interactive system based on self-powered optical communication
24	57	Cao XF, 2021, ACS APPL MATER INTER(13)	3D Printing Magnetic Actuators for Biomimetic Applications
22	57	Sang M, 2021, CHEM ENG J(414)	Flexible PTFE/MXene/PI soft electrothermal actuator with electromagnetic-interference shielding property
47	56	Zafar MQ, 2020, MET MATER INT(26)	4D Printing: Future Insight in Additive Manufacturing
10	52	Forghani S, 2021, INNOV FOOD SCI EMERG(73)	Electrospun nanofibers as food freshness and time-temperature indicators: A new approach in food intelligent packaging
26	47	Huang S, 2021, SMALL STRUCT(2)	Photodeformable Liquid Crystalline Polymers Containing Functional Additives: Toward Photomanipulatable Intelligent Soft ...
42	44	Liu TZ, 2020, COMPOS SCI TECHNOL(186)	4D printed anisotropic structures with tailored mechanical behaviors and shape memory effects

Most Representative Publications

d_{in}	Times Cited	Publication Ref	Publication Title
61	113	Falahati M, 2020, MATER TODAY(40)	Smart polymers and nanocomposites for 3D and 4D printing
57	10	Tariq A, 2023, ADV ENG MATER(25)	Recent Advances in the Additive Manufacturing of Stimuli-Responsive Soft Polymers
56	9	Khalid MY, 2024, EUR POLYM J(205)	3D printing of magneto-active smart materials for advanced actuators and soft robotics applications
54	31	Zhang JF, 2022, ADV MATER TECHNOL-US(7)	Advances in 4D Printed Shape Memory Polymers: From 3D Printing, Smart Excitation, and Response to Applications
52	16	[illegible] J, 2021, POLYM-PLAST TECH MAT(60)	Progress in 4D printing of stimuli responsive materials
52	3	Farid MI, 2023, INT J ADV MANUF TECH(126)	Research on imminent enlargements of smart materials and structures towards novel 4D printing (4DP: SMs-SSs)
50	25	Momeni F, 2020, ENGINEERING-PRC(6)	Laws of 4D Printing
48	35	[illegible] H, 2022, POLYM COMPOSITE(43)	Additive manufacturing of smart polymeric composites: Literature review and future perspectives
47	56	Zafar MQ, 2020, MET MATER INT(26)	4D Printing: Future Insight in Additive Manufacturing
47	22	Pugliese R, 2022, POLYMERS-BASEL(14)	Artificial Intelligence-Empowered 3D and 4D Printing Technologies toward Smarter Biomedical Materials and Approaches
47	1	Zeng CJ, 2020, CAILIAO GONGCHENG(48)	Progress in 4D printing of stimulus-responsive composites and its applications
42	44	Liu TZ, 2020, COMPOS SCI TECHNOL(186)	4D printed anisotropic structures with tailored mechanical behaviors and shape memory effects
42	8	Jiang Z, 2021, MATER ADV(2)	3D and 4D printable dual cross-linked polymers with high strength and humidity-triggered reversible actuation
41	2	[illegible], 2023, GREEN MATER()	Sustainable feedstocks for 4D printing: biodegradable polymers and natural resources

Cluster 10 ("FABRICATION"). Most cited and representative authors. For each author, we display the number N_a of publications that authored in that cluster, the sum TC_a of their number of citations (according to Web Of Science), and the sum k_a of their in-degree.

Most Cited Authors

Author	N_a	TC_a	k_a
Liu XH	3	381	48
Tian MW	3	381	48
Li X	2	339	22
Yang Q	2	337	6
Zhou HM	2	336	5
Chen M	2	336	6
Zeng SN	1	336	3
Pian SJ	1	336	3
Su MY	1	336	3
Wang ZN	1	336	3
Wu MQ	1	336	3
Chen MY	1	336	3
Xiang YZ	1	336	3
Wu JW	1	336	3
Zhang MN	1	336	3
Cen QQ	1	336	3
Tang YW	1	336	3
Zhou XH	1	336	3
Huang ZH	1	336	3

Most Representative Authors

Author	N_a	TC_a	k_a
Liu LW	4	122	132
Leng JS	4	122	132
Tariq A	2	19	113
Arif ZU	2	19	113
Khalid MY	2	19	113
Hossain M	2	19	113
Umer R	2	19	113
Ramakrishna S	3	42	106
Liu YJ	3	52	104
Zeng CJ	2	45	89
Kandasubramanian B	2	36	82
Huang YL	3	58	68
Falahati M	1	113	61
Ahmadvand P	1	113	61
Safaee S	1	113	61
Chang YC	1	113	61
Lyu ZY	1	113	61
Chen R	1	113	61
Li L	1	113	61

FIGURE 2.11 Cluster 10 bibliometric analysis report.

have had an immense effect on the Clusters 3 and 6, Internet of Things and Industry 4.0, respectively.

2.4.4.2 Most Cited Publications

With the entire set of literature available, we highlight the influential contributors in Cluster 1. For instance, in Cluster 1 (CNN), the highest number of citations are for LeCun et al.'s (2015) article that presented the "Deep Learning"-based implementation in the foundation for intelligent manufacturing.

The in-depth scrutiny delivers a thorough synopsis of studies that have been undertaken in the domain of intelligent manufacturing. Through our research, we gained valuable insights into the evolution of the field. Further, we covered the cluster feature, research collaboration trends, current topics, and impact measurements to uncover these findings. The findings reveal valuable insights for future research directions and highlight the importance of collaboration across different disciplines to foster innovation in this domain.

2.5 DISCUSSION

2.5.1 Overview of Key Findings

The bibliometric study provides helpful perspectives into the prevailing state of studies on intelligent as well as smart manufacturing. This section discusses the relevance of the results we obtained, highlighting significant patterns, areas where further research is needed, and potential future directions for the field.

2.5.2 Trends and Patterns

The trends and patterns are studied in the different clusters. Further, we explored significant rises in research in publications, indicating the industry's rising interest in employing digital solutions to enhance production processes. The results highlight significant developments and research trends that are driving innovation in smart manufacturing. The amalgamation of the IoT, AI, machine learning, and neural networks, as well as advanced cyber solutions is reflected in the development of the new framework from Industry 4.0 perspectives. Furthermore, our findings emphasise the multidisciplinary character of intelligent manufacturing research, with cooperation spanning institutions, regions, and research fields. This multidisciplinary approach is visible in clusters such as Technologies and Services/Applications, where researchers from various experiences convene to resolve complicated manufacturing issues.

2.5.3 Research Gaps and Challenges

However, within the discipline of smart manufacturing research, there are, nevertheless, substantial gaps and challenges, irrespective of the gains previously accomplished. Addressing these gaps is paramount to ensuring comprehensive exploration and inclusion of all pertinent technologies and methodologies in future research endeavours. Technologies like blockchain, AR/VR, metaverse, IoT, and cyber physical systems are largely unexplored in the current literature and exhibit huge potential for growth. This divergence reflects the need for cross-border collaboration and information exchange in order to cultivate a research community that is more inclusive and internationally representative. Consequently, research reveals how there are discrepancies in region-specific investments in the area of intelligent manufacturing research, with certain regions producing a greater number of publications than others.

2.6 OPEN ISSUES AND CHALLENGES

To reach their full potential, smart and intelligent manufacturing systems must be deployed and improved. Figure 2.12 illustrates the open topics and challenges.

2.6.1 UNEVEN ADOPTION OF TECHNOLOGIES

Global competitive landscapes can be impacted significantly by technology adoption gaps. When compared with larger corporations, small- and medium-sized businesses

FIGURE 2.12 Breakdown of the outstanding concerns and challenges in intelligent manufacturing.

(SMEs) may have difficulty implementing smart manufacturing systems. There are enormous strategies progressing at an exponential pace to address the challenges in the implementation phase. Further, we need government support in developing smart industries aimed at overcoming these challenges, including good incentives policies, public-private partnerships, and education programs designed to democratise access to smart technologies and provide all players with the necessary skills and resources.

2.6.2 LACK OF STANDARDISED FRAMEWORKS

There is a strong urge to strengthen industrial standards in integrated smart manufacturing systems with emerging technologies. There is a lack of standard protocols to handle IoT devices, heterogeneous data formats, and interoperability in various communication protocols, making this challenge more complicated. Thus, regulatory bodies with complete ethical and environmental standards need to be synchronised in the smart industry. Interoperability and compliance are encouraged through international collaboration and the development of open-source platforms and tools.

2.6.3 SECURITY AND PRIVACY CONCERNS

Manufacturing systems lack the advancement and rising cybersecurity measures tailored to their unique needs based on ongoing research, development, and implementation. The unsecured environment encounters the increasingly high risk of cyber-attacks in interconnected and interoperability and suffers from data breaches. Hence, maintaining secure systems against evolving threats is a challenging concern. Furthermore, organizations must develop a culture of security and continually train their workforce on cybersecurity best practices. Similar work on security using blockchain technologies and its relevant application areas can be referred to in [12,13].

2.6.4 HUMAN-CENTRIC CHALLENGES

The adaptability of the increasing role of AI in the smart industry is also a challenge that leads to job displacement, mismatched skill sets, and issues with human-machine interaction. So, promoter awareness and role-based separation are key to resolving these issues. Developing a range of smart technologies that are effortless for mankind to use, initiatives to help job applicants improve their abilities, and an environment fostering ingenuity and new ideas are all crucial to attaining that goal. Similar work on the adoption of AI can be referred to in [14,15].

2.6.5 SUPPLY CHAIN COMPLEXITY

The global supply chain is gritty and complex in meeting the growing population demands in industries like healthcare, agriculture, and the consumer base. For instance, in the COVID-19 pandemic era, vulnerabilities like reliance on single sources and lack of visibility were exposed. Innovative solutions like blockchain technology and analytics for prediction can help supply chains become more robust, transparent, and diverse.

2.6.6 SUSTAINABILITY AND ENVIRONMENTAL IMPACT

The adoption of a sustainable footprint is a major concern in the current industry to reduce environmental stress. The integration of sustainable, more advanced systems to implement smart manufacturing processes is a challenging aspect to achieve without compromising efficiency or profitability. Further, these initiatives include the need for more optimisation solutions for resource management and minimising waste through the implementation of circular economy principles and the adoption of clean energy sources. Further reading on sustainability and its implication on digital technologies can be found in [16,17].

2.6.7 REGULATORY AND COMPLIANCE ISSUES

Industry 4.0 has witnessed immense ecosystem growth that demands diverse solutions with adaptable strategies to all the challenges. Fostering these technological advancements, instituting regulatory adjustments, and developing educational initiatives are required. This will require integrating these diverse solutions for resilience and innovation in implementing Industry 4.0. Thus, collaboration between stakeholders across the manufacturing ecosystem is key for sustainable, and secure solutions. The legal and regulatory frameworks help maintain a smart manufacturing ecosystem and nurture technology-based advances. Data protection regulations and international standards are complex for manufacturers. Compliance with an ever-changing regulatory environment is the key to ensuring smart manufacturing practices are both efficient and innovative. The implications of ethical AI can be found in [18–20].

2.7 CONCLUSION

According to the results of the comprehensive analysis of 1046 articles on intelligent and smart manufacturing, the field is characterised by a wide range of research areas and an evolution of technology. The fabrication techniques in Industry 4.0 demonstrate a range of advancements with the implementation of convolutional neural networks (CNNs) and the Internet of Things (IoT). This chapter highlighted the multifaceted challenges in integrating smart manufacturing within domains including computer science, engineering, data analytics, and other commercial industries. This chapter presented fresh concepts and innovative approaches for managing remote maintenance and real-time analysis, aiding in their incorporation, while older clusters like IoT reflect established research. Although international collaboration and resource sharing are needed to leverage diverse expertise, the global contribution to the field is notable. There is evidence of significant advancement in research in smart industries, with some clusters showing high levels of citations and h-indices, indicating substantial academic and industrial influence, while others indicate emerging sectors with potential for growth. Moreover, the study emphasizes the need for standardised frameworks, robust security measures, and uneven adoption of technologies. Future research is likely to focus on improving system integration, securing data privacy, promoting sustainable manufacturing, and adapting the workforce to rapidly evolving technologies. Ultimately, intelligent and smart manufacturing are at a pivotal point, with ongoing innovation central to overcoming hurdles and moving towards a more sustainable, intelligent, and efficient future.

REFERENCES

1. T. S. Jones and R. C. Richey, "Rapid prototyping methodology in action: A developmental study," *Educ. Technol. Res. Dev.*, vol. 48, no. 2, pp. 63–80, 2000, https://doi.org/10.1007/BF02313401.
2. K. Kukushkin, Y. Ryabov, and A. Borovkov, "Digital twins: A systematic literature review based on data analysis and topic modeling," *DATA*, vol. 7, no. 12, 2022, https://doi.org/10.3390/data7120173.
3. H. Seki, T. Nose, Y. H. Lee, C.-F. Chien, and M. Gen, "Special issue on recent advance in intelligent manufacturing systems foreword," *Comput. \& Ind. Eng.*, vol. 65, no. 1, p. 1, May 2013, https://doi.org/10.1016/j.cie.2012.07.001.
4. M. Qiao and B. Meng, "Technical construction of intelligent processing production line for aircraft structural parts," in *2019 5th International Conference on Green Power, Materials and Manufacturing Technology and Applications (GPMMTA 2019)*, 2019, vol. 2185, https://doi.org/10.1063/1.5137889.
5. S. Kumar, P. Tiwari, and M. Zymbler, "Internet of things is a revolutionary approach for future technology enhancement: A review," *J. Big Data*, vol. 6, no. 1, 2019, https://doi.org/10.1186/s40537-019-0268-2.
6. O. Pisacane, D. Potena, S. Antomarioni, M. Bevilacqua, F. Emanuele Ciarapica, and C. Diamantini, "Data-driven predictive maintenance policy based on multi-objective optimization approaches for the component repairing problem," *Eng. Optim.*, vol. 53, no. 10, pp. 1752–1771, October 2021, https://doi.org/10.1080/0305215X.2020.1823381.
7. B. Mocanu, R. Tapu, and T. Zaharia, "Utterance level feature aggregation with deep metric learning for speech emotion recognition," *Sensors*, vol. 21, no. 12, 2021, https://doi.org/10.3390/s21124233.
8. C. Cronin, A. Conway, A. Awasthi, and J. Walsh, "Flexible manufacturing using automated material handling and autonomous intelligent vehicles," in *2020 31st Irish Signals and Systems Conference (ISSC)*, 2020, pp. 236–241.
9. Y. Wang, Y. He, P. Zhu, L. Zhang, and F. Zhao, "New processing method based on intelligently manufacturing blade with multiple space and compound angles," in *Communications, Signal Processing, and Systems, CSPS 2018, Vol III: Systems*, 2020, vol. 517, pp. 647–655, https://doi.org/10.1007/978-981-13-6508-9_78.
10. S. Liu, J. Qu, and J. Pan, "Research of enterprise agile intelligence manufacture technique," in *CEIS 2011*, 2011, vol. 15, https://doi.org/10.1016/j.proeng.2011.08.554.
11. J. Yu, M. Feng, P. Huang, and Q. Zhang, "Research on design and development for intelligent manufacturing execution system of tungsten powder processing," in *Frontiers of Manufacturing and Design Science II, PTS 1–6*, 2012, vol. 121–126, pp. 4080–4084, https://doi.org/10.4028/www.scientific.net/AMM.121-126.4080.
12. T. Hai, et al. A Novel & Innovative Blockchain-Empowered Federated Learning Approach for Secure Data Sharing in Smart City Applications. In: Iwendi, C., Boulouard, Z., Kryvinska, N. (eds) *Proceedings of ICACTCE'23 — The International*

Conference on Advances in Communication Technology and Computer Engineering. ICACTCE 2023. Lecture Notes in Networks and Systems, vol 735. Springer, Cham, 2023, https://doi.org/10.1007/978-3-031-37164-6_9

13. A. Raj, A. Kumar, V. Sharma, S. Rani, A. K. Shanu, and T. Singh, "Applications of Genetic Algorithm with Integrated Machine Learning," *2023 3rd International Conference on Innovative Practices in Technology and Management (ICIPTM)*, Uttar Pradesh, India, 2023, pp. 1–6, https://doi.org/10.1109/ICIPTM57143.2023.10118328.

14. M. A. Ali, B. Balamurugan, R. K. Dhanaraj, and V. Sharma, "IoT and Blockchain based Smart Agriculture Monitoring and Intelligence Security System," *2022 3rd International Conference on Computation, Automation and Knowledge Management (ICCAKM)*, Dubai, United Arab Emirates, 2022, pp. 1–7, https://doi.org/10.1109/ICCAKM54721.2022.9990243.

15. A. Raj, A. Kumar, V. Sharma, S. Rani, A. K. Shanu, and H. K. Bhardwaj, "Decipherable for Artificial Intelligence in Medicare: A Review," *2023 4th International Conference on Intelligent Engineering and Management (ICIEM)*, London, United Kingdom, 2023, pp. 1–6, https://doi.org/10.1109/ICIEM59379.2023.10165690.

16. P. Anamika, B. Balusamy, V. Sharma, "Introduction," in *Disruptive Artificial Intelligence and Sustainable Human Resource Management: Impacts and Innovations – The Future of HR*, River Publishers, 2023, pp. 1–14.

17. P. Ajmani, V. Sharma, S. Sharma, A. Alkhayyat, T. Seetharaman, and Z. Boulouard, "Impact of AI in financial technology- A comprehensive study and analysis," *2023 6th International Conference on Contemporary Computing and Informatics (IC3I)*, Gautam Buddha Nagar, India, 2023, pp. 985–991, https://doi.org/10.1109/IC3I59117.2023.10398111.

18. V. Rajasekar, K. Venu, V. Sharma, and M. Saracevic, Algorithmic Strategies for Solving Complex Problems in Financial Cryptography. In: Seethalakshmi, V., Dhanaraj, R. K., Suganyadevi, S., Ouaissa, M. (eds) *Homomorphic Encryption for Financial Cryptography*. Springer, Cham, 2023. https://doi.org/10.1007/978-3-031-35535-6_10.

19. V. Sharma, B. Balusamy, M. Sabharwal, and M. Ouaissa (eds.). *Sustainable Digital Technologies: Trends, Impacts, and Assessments* (1st ed.). CRC Press, 2023. https://doi.org/10.1201/9781003348313.

20. S. Ahmad, S. Mishra, and V. Sharma, Green Computing for Sustainable Future Technologies and Its Applications. In: Grima, S., Sood, K., Özen, E. (ed.) *Contemporary Studies of Risks in Emerging Technology, Part A (Emerald Studies in Finance, Insurance, and Risk Management)*, Emerald Publishing Limited, Bingley, pp. 241–256. https://doi.org/10.1108/978-1-80455-562-020231016.

3 Amalgamation of Disruptive Technologies for Implementation of Intelligent Manufacturing

Nidhi Agarwal, Archana Singh, and Mary Cade T. Ambojia

3.1 INTRODUCTION

The rapid progress in the use of cutting-edge technology in intelligent manufacturing has the potential to completely transform the process of designing, producing, and distributing items [1]. This field is advanced by a multitude of innovative technologies [2]. Organizations are progressively embracing these advancements to enhance manufacturing operations, enhance efficiency, reduce expenses, and sustain a competitive advantage in today's always changing commercial market [3].

The Fourth Industrial Revolution, also called Industry 4.0, is characterised by the extensive implementation of advanced technical breakthroughs in the industrial industry. This methodology, alternatively referred to as intelligent and smart manufacturing, utilises automation and digitisation to optimise production operations and augment product functionality [4]. It utilises cutting-edge informational and production technology to facilitate flexible and effective manufacturing processes. Smart factories are essential to intelligent manufacturing as they include advanced technologies such as large amounts of data, IoT, and cloud computing to function autonomously, resulting in enhanced productivity and adaptability [5,6].

Intelligent manufacturing utilises a combination of large amounts of data, CPS, cloud technologies, Internet of Things (IoT), and Machine to Machine (M2M) connections [7]. Additionally, it leverages preexisting relationships and the expertise of human employees. The German government spearheads the implementation of Industry 4.0 to enhance competitiveness and promote the progress of factory automation [8]. The objective of this strategy is to enhance productivity and decrease expenses by implementing efficient data exchange and managing production resources effectively.

Nevertheless, the implementation of intelligent manufacturing systems entails other obstacles, such as the requirement for uninterrupted power supply, efficient

DOI: 10.1201/9781032630748-3

usage of the labor force, and a complex technological infrastructure. Smart factories prioritise automation, minimizing the need for human labor, which may have a negative impact on employment [9]. Intelligent manufacturing overcomes problems and offers various advantages, including enhanced productivity, increased operational flexibility, and the opportunity to provide consumers with more useful products and services.

Embracing Industry 4.0 essentially involves adopting cutting-edge technologies and updating traditional production techniques. Through the use of technology like big data as well as IoT, Industry 4.0 seeks to accomplish the goals of flexible and efficient manufacturing. Intelligent manufacturing deals with increasing efficiency as well as competitiveness in the manufacturing business [10], but it also presents problems and may have a negative effect on production. Intelligent manufacturing can make use of several emerging technologies, including those listed below.

- The IoT allows for the real-time collection and exchange of data from any internet-connected device, sensor, or machine. This information can be used to track production, enhance quality, and cut expenses. Sensors and gadgets combined with IoT may be built into different types of machinery, tools, and consumer goods to provide continuous monitoring and feedback. Uses for this information include preventative maintenance, quality assurance, and monitoring production rates.
- The use of AI and ML allows for the optimisation of production processes, the identification of abnormalities, and the implementation of decisions in real time based on the massive volumes related to data created through IoT devices. Predicting machine faults and recommending maintenance schedules are also possible thanks to machine learning algorithms. The term "artificial intelligence" is defined as a computer's ability to learn and solve problems without direct human involvement. This information can be utilised to fine-tune machinery, foresee when machines will break down, and create brand-new items.
- The manufacturing process generates a large amount of data, which may be processed and analysed using big data analytics to shed light on potential improvement areas, quality problems, and production bottlenecks. Additive manufacturing, for example, 3D printing, enables the efficient and cost-effective production of a wide range of complicated items. It also works well for making prototypes rapidly.
- The usage of AR and VR technologies in education, upkeep, and remote assistance are all possible. They can help assemblers with complicated tasks and give them access to information in real time. The immersive learning experiences made possible by AR (augmented reality) and VR (virtual reality) are changing the face of education. Augmented reality (AR) improves the learning experience in scientific topics by superimposing digital information onto the real world. On the other hand, virtual reality (VR) fully immerses learners inside a simulated environment, allowing them to vividly experience historical events within the confines of the classroom. Both

technologies have demonstrated efficacy in enhancing student engagement, enhancing memory retention, and promoting learning via interactive experiences and practice.

- Blockchain technology provides a reliable and clear method for overseeing supply chains, allowing for the monitoring of both raw materials and completed goods. This technology guarantees the capacity to track items and preserve their condition by generating an immutable log of transactions, promoting transparency, and establishing trust among all parties involved. Implementing robust security measures, such as sophisticated authentication and encryption protocols, safeguards data against manipulation and significantly mitigates the risk of fraudulent activities, hence establishing a highly dependable supply chain environment.

- Simultaneously, cutting-edge robotics and automation assume hazardous or labor-intensive duties, liberating humans from the necessity of carrying them out. These advanced systems, powered by state-of-the-art AI and machine learning technology, are revolutionizing various industries by raising operational efficiency, minimizing the need for human labor, and optimizing accuracy. They facilitate the accomplishment of perilous or difficult jobs, expanding the scope of potential activities in several industries, including manufacturing, logistics, and healthcare.

- Additive manufacturing, also known as 3D printing, enables the quick creation of prototypes and personalised products, significantly decreasing production time and enabling manufacturing as needed. This method expedites the production of tangible models or duplicates of product designs by utilising computer-aided design (CAD) software and 3D printing for swift prototyping in the course of product development. Engineers and designers gain advantages from the capability to rapidly evaluate and improve their ideas, resulting in time and resource savings during the product creation procedure by facilitating quicker iterations and design modifications. Rapid prototyping speeds up the innovation process and shortens the time it takes to get a product to consumers.

- Cybersecurity denotes methods used by preventing, detecting, as well as responding to cyber threats. Firewalls, encryption, threat detection, and user authentication are only some of the technologies, processes, and practises that make up cybersecurity. Cybersecurity is essential in today's interconnected world to ensure the safety of private data and the continued reliability of online services. Strong cybersecurity measures are necessary to secure sensitive data and prevent cyberattacks as manufacturing systems grow more linked.

- In order to reduce latency and allow for quicker decision-making, edge computing moves data processing and analytics closer to the data source (such as IoT devices). Internet-of-Things (IoT) devices are everyday items that have been retrofitted with electronic circuitry, networking protocols, and processing power so they can gather and transmit data across a network. These gadgets are used for a broad range of related purposes, including environmental monitoring and smart home automation with healthcare (through fitness trackers) and manufacturing (via sensors). IoT devices facilitate automation, data analysis, and remote control, which although revolutionary, also present security and privacy concerns for several sectors.

- Manufacturing data can be stored and analysed in the cloud, where online services can promote collaboration and offer scalability.
- Sensors and actuators: State-of-the-art sensors and actuators allow for precise monitoring and regulation of environmental conditions including temperature, humidity, pressure, and motion throughout the production process. These are instruments which can search as well as measure transformation considering environmental situations, such as heat, pressure, light, humidity, and motion. These physical world inputs are transformed into electrical impulses or digital data by sensors. Thermostats, cell phones, and smart homes all use sensors, such as temperature and proximity detectors. In contrast, actuators carry out operations in response to commands from sensors or control systems. They are the ones who actually make things happen, like moving or controlling machinery. Actuators come in many forms, but the most common ones include motors, servos, solenoids, and hydraulic and pneumatic systems. Robotics, factory automation, and other fields can all benefit from actuator-enabled automation.
- Connectivity at the 5G level: Super-fast 5G networks allow for instantaneous data transfer between machines in smart factories. Fifth-generation (5G) connectivity describes the latest iteration of wireless communication standards. The data transfer speeds, latency, and device capacity of 5G networks are vastly superior to those of 4G, 3G, etc. networks. There are many uses made possible by such high-level communication, like IoT, AR/VR, as well as remote surgery. The advent of 5G networks holds great potential for transforming the way networks connect and engage with digital technologies, leading to exciting new opportunities for businesses of all sizes and superior user experiences.
- The term "digital twin" refers to a digital copy of a real-world product, service, or system. They make it possible to model, enhance, and keep tabs on physical systems and processes in the real world. The digital twin has an electronic representation related to the physical system as well as a component which is generated in real time using digital tools and data. In fields as diverse as manufacturing, healthcare, and city planning, it offers real-time monitoring, modelling, and analysis to aid in decision-making, predictive maintenance, and optimisation.
- Technologies such as cobots (collaborative robots) facilitate human-machine collaboration, bringing together the best of both worlds. Advanced automation systems meant to work in tandem with humans in open environments are known as collaborative robots, or cobots. Unlike conventional industrial robots, cobots are designed with safety in mind; they have sensors and software that can detect and respond to the presence of humans, allowing for risk-free teamwork. They assist workers with a broad range of

activities related to industries, production, as well as assembly considering logistics including healthcare, raising output, and decreasing worker danger. Because of their adaptability, re-programmability, and low price, cobots are becoming increasingly popular in a variety of sectors and among SMEs for the purpose of automating and streamlining processes while still allowing for human-machine collaboration.

The concept of Industry 4.0 (I4.0) represents a new stage of technological development in manufacturing. Industry 4.0 aims to enhance production effectiveness and efficiency through the integration of advanced technologies [11]. It involves the use of the Internet of Things (IoT) in manufacturing shop floors and management activities like logistics and planning. This integration creates a global network where parties can control and exchange information. Industry 4.0 also impacts production systems by optimising allocation regarding various resources, minimising human sources including the costing factor of logistics, and enhancing flexibility related to business processes [12,13,14,15,16]. The effective use of technologies like ERP (Enterprise Resource Planning), DCSs (Distributed Control Systems), and MESs (Manufacturing Execution Systems) involves collecting and analyzing data to control machinery and monitor condition. Automation plays a critical part in achieving the goals of Industry 4.0, such as improving efficiency, reducing maintenance events, and enabling predictive maintenance.

3.2 IMPACT OF INDUSTRY 4.0 CHARACTERISTICS ON MAINTENANCE MANAGEMENT SYSTEMS

MMS have an effect on the characteristics of I4.0. Digital device and technology integration allow maintenance workers to get instant access to equipment paperwork and service history right when it's needed. They may now more quickly resolve difficulties and spend less time looking for answers online [17,18]. Industry 4.0 (I4.0) is symbolised by the increased use of automation, which includes digitisation in manufacturing and related sectors. Maintenance management systems (MMS) are influenced by I4.0 components like IoT, evaluation of big data, AI, robotics, and automation.

In the following analysis, the specific effects that each component of Industry 4.0 has on MMS are explored in greater detail.

- Predictive maintenance employs data analysis including artificial intelligence to anticipate mechanical malfunctions in advance, thereby minimising operational interruptions and prolonging the lifespan of assets, including optimising the allocation of maintenance resources.
- IoT allows for the immediate gathering of data from various sources, including the current condition and functioning of machines. This abundance of information contributes to the improvement of predictive maintenance, optimises maintenance schedules, and expedites the early identification and resolution of problems.

- Big data analytics entails the scrutiny and manipulation of enormous quantities of data gathered from machinery and sensors. This research offers useful insights into the health and performance of equipment, resulting in benefits such as enhanced predictive maintenance, optimum scheduling of maintenance, and increased durability of assets.
- Artificial intelligence (AI) is essential for the development of predictive maintenance models, as it enables the analysis of data from equipment and sensors. AI also automates maintenance choices, resulting in a more efficient maintenance process and improved operational efficiency. This has the potential to increase predictive maintenance's precision and efficiency while decreasing the requirement for manual intervention.
- Robotics and automated systems can be programmed to carry out mundane upkeep duties like oiling, inspecting, and cleaning. As a result, maintenance workers will have more time for higher-level activities and fewer opportunities for human error.

Machines can now self-diagnose problems and schedule repairs thanks to Industry 4.0. As a result of being able to do routine maintenance without human intervention, maintenance management systems are more effective as a whole [19,20]. Industry 4.0 improves the efficiency of maintenance management systems by making data more easily accessible, allowing for preventative upkeep, and maximising the use of available resources. Among the most significant effects are:

- By collecting and analysing data in real time from machinery and equipment, I4.0 technologies like IoT, evaluation of big data, including AI, can improve predictive maintenance. Preventive maintenance can be conducted and costly downtime avoided if problems are detected using this data in advance of their occurrence.
- Maintenance expenses can be lowered through the use of I4.0 functions by enhancing predictive kinds of maintenance and decreasing downtime.
- Maintenance activities can be created more efficiently through the use of I4.0 technologies. For instance, augmented reality (AR) can be utilised to give maintenance workers guidance from afar and eliminate the need for manual instructions.
- Features of Industry 4.0 can aid organisations in better managing their assets. Digital twins, for instance, can be used to model real-world possessions in a digital environment. Asset status monitoring, improved maintenance planning, and replacement preparation can all benefit from this data.

3.2.1 Characterisation of the New Innovation Which Resolves Major Issues

3.2.1.1 Automation

Industry 4.0's improved manufacturing process rests in large part on the back of automation. Manufacturing efficiency, adaptability, and output can all be improved through the use of various methods that fall under the umbrella of "automation," which refers to I4.0. Major areas related to automation in Industry 4.0 manufacturing processes are made possible by the widespread adoption of various current technologies such as IoT, including AI with evaluation of big data, and cloud computing. It should be noted that I4.0 depends largely upon automation since this allows for the use of fully autonomous technologies that can make choices and even eliminate human labor altogether. KUKA's intelligent robots and M2M connectivity are just two examples of the automation technologies used in the company's "smart factories," which maximise output and efficiency. Benefits such as reduced risk of injury, increased productivity, and enhanced competitiveness are realised as a result of the automation made possible by these technologies. By sharing data and coordinating actions, the automated factories of Industry 4.0 hope to reduce waste and cut costs without sacrificing productivity.

The KUKA Smart Factories are a suite of automation services available to several sectors. They can move aeroplane parts with millimetre accuracy using the KUKA Omni Move movable transport system. The omnidirectional Mecanum wheels on this platform make it possible to transport massive objects. KUKA also employs M2M techniques and smart robots for machine tool automation, thereby complementing Industry 4.0 principles. KUKA's Augsburg facility is a standard example of a global machine manufacturer's production environment, with seven robots at work. Medical high-tech items are just some of the many that KUKA offers as part of its automation solutions for the medical sector.

1. Industry 4.0's central idea is the creation of highly linked, smart factories that use real-time data from a variety of sources to optimise operations. Included in this category are sensors, Internet of Things gadgets, and factory automation setups.
2. Integrating IoT and sensors considering manufacturing process allows for the collection of data on a broad range of characteristics, including temperature, humidity, machine health, and product quality. Monitoring in real-time, predictive maintenance, and process optimisation all rely on this data.
3. Sophisticated robots are employed for several applications, including material handling, assembly, quality control, and even human-robot collaboration (cobots). They are flexible in their approach to work and the jobs they perform.
4. Additive manufacturing, which includes 3D Printing), enables speedy prototype, individualised design, and even mass manufacture of complicated components and goods. It aids in the prevention of waste and the implementation of greener production techniques.
5. Predictive maintenance, quality control, demand forecasting, and process optimisation are among the areas where AI algorithms have found use. Machine learning algorithms can examine massive information, find patterns within them, and then act accordingly in real time.
6. Machine learning algorithims allow for continuous tracking and simulation, which improves management, optimisation, and preventative care.

7. Supply chain optimisation, Automation moves out of the factory and throughout the whole process.

The KUKA Omni Move mobile transport platform [21] makes this possible by providing a means of transporting big loads in any direction.

Similarly, utilising KUKA's intelligent robot applications and M2M technology, loading and unloading machines are just two examples of how the robot-based KUKA system technology using machine tool automation supports parts of Industry 4.0 [22]. Seven robots are used in production at the KUKA site in Augsburg, which is normal for an international machine company [23]. KUKA robots are well-suited via a wide range utilising medical technology applications and represent yet other applications related to I4.0 in the medical area, along with automation applications for increased efficiency regarding hospitals. These solutions can be used in any field from diagnostics and surgery to therapy. To that end, KUKA provides a vast selection of cutting-edge medical technology, including anything from surgical robots to diagnostic and therapeutic aids [24,25].

3.2.1.1.1 Kano model
Businesses who are interested in improving their products and services can learn a lot by applying the Kano model to automation solutions. This approach can classify automation features and evaluate their effect on client happiness. The Kano model aims to tailor automation solutions in Industry 4.0 to address varying customer needs effectively, ensuring a more customer-centric approach to innovation and technology implementation (see Table 3.1).

3.2.1.2 Interconnection
3.2.1.1.1 Connectivity Improvements in Industry 4.0
The incorporation of cutting-edge instruments and technology into Industry 4.0 results in vastly enhanced connectivity. Because of this linkage, stakeholders in the supply chain may easily work together, integrate their systems, and share information with one another. Cyber-physical production systems (CPPS) can swiftly adjust to fluctuations in demand, broken machinery, or a lack of inventory thanks to vertical networking. It also helps conserve resources and decreases unnecessary waste. The full promise of the Fourth Industrial Revolution can only be realised with enhanced connectivity in Industry 4.0. Automation, artificial intelligence, and big data analytics play prominent roles in the Fourth Industrial Revolution, which is characterised by the merging of digital, physical, and biological technology. In order to work, each of these technologies necessitates a connection that is both stable and fast.

In Industry 4.0, connections can be enhanced in a variety of ways. The rollout of 5G networks represents a significant strategy. Faster data transfer rates, lower latency, and higher reliability are just a few of the improvements that 5G cellular networks bring over their predecessors. Among these are:

TABLE 3.1

KANO Model to Assess the Innovation of Automation in Industry 4.0

Feature	Innovation	Need Type	Priority
Data-driven solution	– Development as well as evaluation of various services like Data Acquisition with Programmable Logic Controller, including PID Controller – Development & analysis of different services like Data Acquisition with Programmable Logic Controller, including PID Controller – Automation of product assessment using Recycling Model of Product – Dedicated module operating using CAD 3D kind of system environment – Cyber-physical system (CPS) with smart sensors and actuators – Internet of Things (IoT) for sensor technologies and automation	Performance	High
Automation in equipment management	– Use of AutomationML language for Asset Administration Shell – Integration of OPC UA for information and communication layer	Important	Low
Satisfaction at managerial level	– Managers have higher satisfaction and more extensive use of digitalisation experiences compared to operators – Digitalisation and digital tools seem to have only reached the managerial level	Basic	Low
Easy logistics and supply chain solutions	– I4.0 centers around the implementation of automation, digitalisation, and information exchange. The objective is to attain a technologically advanced factory	Important	High
Automation in production development path	– The notion of manufacturing procedure evolution is centered around the utilisation of existing automation technology. This is exemplified by a project in the area of technical fluid maintenance – Artificial intelligence, machine learning, and automated technologies – AMCoT – AVM	Critical	High
Customer service	– Automated chatbots and other tools can be used to provide 24/7 customer support and answer customer questions	Basic	High

(Continued)

TABLE 3.1 (CONTINUED)
KANO Model to Assess the Innovation of Automation in Industry 4.0

Feature	Innovation	Need Type	Priority
Automation in inventory management	– Integration of IoT, AI, and deep learning – Tracking and managing inventory in real-time – Integrative R&D framework for inventory systems modelling and optimisation – Computing devices coupled to a database multi-dimensional model of sales floors and inventory items	Performance	High
Predictive maintenance	– Combinatory process synthesis for automated generation of workflows – Constraint-based variant compilation for generating solution variants	Critical	Medium

- Edge computing involves moving processing power closer to the sources of data and the users, resulting in reduced latency and improved performance for a wide range of Industry 4.0 applications.
- IoT links a wide range of devices that may independently gather and send data. This abundance of data facilitates a more profound comprehension of operations and bolsters more strategic decision-making.
- Industrial ethernet represents a customisecustomised iteration of ethernet designed specifically for the manufacturing industry and related settings. The capabilities of real-time communication as well as time synchronization are crucial for the achievement of Industry 4.0 projects. They provide improved dependability and robustness compared to traditional ethernet.

Horizontal integration facilitates the optimization of processes, increases flexibility, and allows for customisation by promoting communication between different stakeholders. Industry 4.0 enhances efficiency, productivity, and adaptability to market changes by improving connection. Industry 4.0, also referred to as Industrial IoT, greatly enhances the networking possibilities of manufacturing companies.

1. 5G with Wi-Fi Coexistence: The research emphasises that 5G will function concurrently with Wi-Fi and conventional cable connections in industrial settings. This allows for ultra-low latency, high reliability, and support for a large number of devices with various traffic types.
2. Benefits of 5G: 5G enables true mobility for all devices, saves on wiring costs, and fosters automation in the manufacturing industry. It offers seamless connectivity for automated guided vehicles and allows for real-time process control through targeted sensor-driven analyses.

3. Complexity in the Short Term: The deployment of 5G private networks may increase the complexity of operations in the short term. Manufacturing plants are not accustomed to running cellular networks. However, as deploying and running a 5G private network becomes as easy as Wi-Fi, the complexity is expected to decrease.

4. IIoT Platform: Industry 4.0 utilises contemporary cloud technologies to expand processing capabilities, oversee data management, and utilise applications including visualisation of data, machine learning, and information analytics. Nevertheless, physical interfaces remain necessary for gathering data from the work floor.

5. IIoT Hardware Unit Sales: The projected sales of 5G IoT hardware units for Indistry 4.0 showed a significant rise beginning in 2023. This indicates the increasing adoption and importance of connectivity in the manufacturing space.

6. Digital Innovation: Connectivity improvements in Industry 4.0 enable manufacturers to leverage core technologies and implement a wide range of use cases. These use cases include digital performance management, remote monitoring and control, condition-based maintenance, energy optimization, advanced automation, and more.

7. Value Creation: By embracing digital innovation and leveraging connectivity, manufacturing companies can achieve substantial efficiency gains and drive innovation. Integration as well as ecosystems need manufacturers by joining in and catching up to stay relevant in the industry.

3.2.1.3 Sustainability

The sustainability of Industry 4.0 manufacturing processes varies depending on the specific technologies and practices being used. Responsible practises are essential to ensuring the long-term viability of Industry 4.0 [26], which is characterised by automation, IoT, and data-driven operations. It may improve things like resource productivity, trash minimisation, and workflow optimisation. This includes the following:

1. Enhanced Productivity: Technologies derived from Industry 4.0 have the capability to boost the efficiency of manufacturing operations, leading to reduced energy consumption and less waste production.

2. Waste Minimisation: By facilitating improved production planning and quality management, Industry 4.0 technologies can contribute to significant waste reduction.

3. Adoption of Renewable Energy: Manufacturing locations that use Industry 4.0 can effectively employ renewable energy sources such as solar and wind to reduce carbon emissions.

Based on research done by the World Economic Forum, Industry 4.0 has the capacity to decrease worldwide greenhouse gas emissions by as much as 30%. Additionally, the study revealed that Industry 4.0 has the potential to generate a staggering 20 million new employment opportunities on a global scale. This has created an extensive factor model to evaluate and encourage sustainability in the context of Industry 4.0. This model encompasses multiple aspects and essential variables that contribute to the implementation of sustainable practices in industrial processes. It includes domains like energy efficiency, resource allocation, minimisation of environmental effect, adherence to ethical and legal standards, and promotion of labor welfare. Each of these criteria is further subdivided into particular indicators, enabling a detailed assessment of sustainability initiatives. By employing this factor approach, organisations can obtain a comprehensive perspective for their sustainability performance within Industry 4.0, pinpoint areas that need enhancement, and execute strategies to promote enduring social and environmental consciousness while upholding operational efficiency as well as competitiveness.

3.2.1.3.1 Factor Analysis

The research examines the intricate challenges that emerge when adjusting to the constantly evolving realm of state-of-the-art technology in Industry 4.0, with a focus on sustainability. Several factors need to be taken into account in this context. One significant barrier to achieving sustainability is the unavoidable rise in energy usage that comes with the implementation of Industry 4.0 technologies. The rapid obsolescence of technological components exacerbates this issue by generating a substantial amount of electronic waste. The necessity for stringent regulations has arisen due to concerns around data privacy and security, including monitoring, hence shifting the emphasis towards the ethical aspect. Industry 4.0 holds the capacity to enhance sustainability through various means, such as optimising resource utilisation, reducing waste, and improving supply chain management. The table of factor analysis (Table 3.2) suggests that finding a balance between leveraging the benefits presented by Industry 4.0 and acknowledging the challenges it brings for long-term sustainability is crucial.

Industry 4.0, often known as the Fourth Industrial Revolution, is characterised by the merging of cyber, physical, and biological systems. The technology possesses numerous potential applications, although it also faces certain obstacles that must be addressed to ensure its long-term sustainability.

- Industry 4.0 tools face a significant challenge due to their substantial energy demands. Operating facilities such as data centers can incur significant expenses. The manufacturing and utilisation of electronic devices also require significant quantities of electricity.
- One additional challenge is the adverse impact which Industry 4.0 technology may have on the environment. Hazardous waste is generated during the production of electronic devices. Moreover, the utilisation of these devices might significantly contribute to environmental pollution along with other ecological concerns.
- Industry 4.0 technology also brings up several socioeconomic challenges. A particular cause of concern is the possibility for job losses resulting from automation. The data gathered through Industry 4.0 technologies gives rise to more concerns regarding privacy and security (Table 3.2).

TABLE 3.2

Factor Analysis: Sustainability Challenges of Industry 4.0

Sustainability Challenge	Key Dimensions	Description
Energy Consumption	1. Energy Efficiency of Technology	The conservation of energy of Industry 4.0 technologies & processes, such as Internet of Things (IoT) devices and intelligent manufacturing systems.
	2. Renewable Energy Adoption	The utilisation of renewable energy sources in powering Industry 4.0 processes.
Resource Management	1. Raw Material Usage	Optimising the utilisation of raw materials through practices such as recycling and minimising waste.
	2. Water Usage	Efficient and environmentally friendly water management in the context of Industry 4.0.
	3. Resource Recycling	The practice of recycling and reusing components and materials is employed to mitigate the depletion of resources.
Environmental Impact	1. Emissions Reduction	Efforts to mitigate greenhouse gas emissions while decreasing air pollution.
	2. Waste Management	Efficient and effective handling and control of waste produced as a result of Industry 4.0 operations.
Sustainable Supply Chain	1. Supply Chain Transparency	Implementing measures to guarantee openness and accountability within the supply chain in order to tackle concerns such as ethical sourcing and equitable labor practices.
	2. Resilience and Risk Management	Overseeing the mitigation of supply chain threats and disruptions including vulnerabilities, which include addressing cybersecurity threats.
	3. Localised Production	Advocating for the implementation of localised as well as dispersed production methods in order to mitigate emissions associated with transportation.
Workforce Impact	1. Skills Development	Enhancing the skills and knowledge of the workforce to effectively adjust to emerging technologies and job responsibilities.
	2. Job Displacement Concerns	Addressing worries around automation and the potential displacement of jobs.
	3. Health and Well-being	Promoting the physical and mental welfare of employees in highly advanced technological settings.
Ethical and Legal Issues	1. Data Privacy and Security	Ensuring the confidentiality and integrity of data gathered and utilised in Industry 4.0 systems.
	2. Intellectual Property Protection	Protecting intellectual property rights during the digital era.
	3. Regulatory Compliance	Ensuring adherence to pertinent sustainability and business regulations.
	4. Ethical AI and Automation	Examining ethical issues associated with artificial intelligence (AI) and automation, including bias and accountability.

Table 3.2 offers a systematic summary of the sustainability issues linked to Industry 4.0 and categorises them into essential aspects for a more thorough examination. It is important to note that the particular difficulties and aspects may differ based on the sector, location, and circumstances under which Industry 4.0 technologies have been utilised. Additional examination and investigation may be required to customise the table according to the specific requirements and difficulties of a certain company.

3.2.1.4 Customisability

The level of customisation has been enhanced in Industry 4.0 as a result of the implementation of digital technologies and increased automation. In the context of Industry 4.0, manufacturing procedures have undergone enhancements to increase their flexibility and adaptability in order to cater to the specific preferences of customers and accommodate the dynamic shifts in market needs. Industry 4.0, also referred to as the Fourth Industrial Revolution, focuses on the capacity to customise products according to specific requirements. This concept embodies the degree to which companies may adapt their products, services, and manufacturing methods to meet the distinct and always evolving needs of their customers. Technological progressions, like the Internet of Things, artificial intelligence, large-scale data analytics, and automation, contribute to enabling customisation within the framework of Industry 4.0 (Table 3.3).

This means that even when goods are produced on a massive scale, they can still be tailored to suit the specific needs of each buyer. The same assembly line can build vehicles with varying options, colours, and specs to meet the needs of a wide range of customers. Customisation has been shown to increase individual demand, willingness to pay, and positive product reviews. The author conducted semistructured interviews with 16 marketing managers at businesses that offer product customisation to see if our earlier findings were consistent with their perspectives. This argues that academics and business leaders have overlooked the risk that customers' perceptions of essential product features could be negatively impacted by the customisation process. If customers helped make a dish, would they have a different impression of how nutritious it was? If they designed their own T-shirt, would they think it was more or less trendy? We hope to find the answers to these issues by studying how

TABLE 3.3
ICE Scoring Model for Industry 4.0

Initiative	Impact	Confidence	Ease	ICE Score
Digital Design and Simulation	5	4	3	60
Develop a New Product using 3D Printing	4	3	2	24
Automate the Order Fulfilment Process	3	5	4	60
Customer-Driven Production	4	4	2	32
Batch and One-Off Production	4	2	3	24

customers' opinions of key product attributes change depending on whether or not they are customised (or not).

Importantly, this assumes that customisers (i.e., customers involved in product modification) modify their product perceptions according to their self-image, whereas product perceptions largely depend on extrinsic cues. This assumption stems from research showing that when people have a hand in creating a product, they are more likely to see parallels between the product and themselves. This can be proposed that an individual's product perceptions will also be influenced by specific traits of his or her self-image while prior studies have shown a transfer of generalised positive self-esteem to products [27,28,29,30]. Existing studies and managers' insights suggest this can lead to more positive product impressions sometimes, but this can also lead to less positive perceptions, depending on the product and the customer's self-image. A consumer who views themselves as an unhealthy eater, for instance, is more likely to regard a customised food option as less healthful than a ready-made one. Self-image-consistent product perceptions is the term used to describe this phenomenon.

The configurable characteristics of Industry 4.0 offer numerous benefits. By utilising this tool, organisations can enhance customer satisfaction by delivering tailored products and services that precisely cater to their individual requirements. Due to the reliance on consumer demand, the likelihood of excessive overproduction or stockpiling is reduced. Moreover, the capacity to customise encourages innovation by stimulating the development of new and unique product variations.

Nevertheless, there are disadvantages associated with the ability to customise in the context of Industry 4.0. These include the requirement for advanced data analytics to understand customers' preferences, the intricacy of manufacturing procedures to accommodate personalisation, and the potential for unauthorised access to sensitive consumer data.

Customising products and services for individual consumers is a crucial aspect of Industry 4.0, a movement that seeks to enhance productivity and innovation in the manufacturing and delivery of goods and services. This improvement is ascribed to:

1. Digital Design and Simulation: Industry 4.0 enables the development of digital twins that are virtual models of goods and procedures. Manufacturers have the ability to digitally model and customise product designs, facilitating the process of adjusting and optimising them to meet specific client requirements.
2. Additive Manufacturing: Additive Manufacturing, often known as 3D Printing, allows for the efficient and quick production of highly personalised components and products at a lower cost. This allows for on-demand manufacturing of unique items with minimal setup and tooling costs.
3. Flexible Automation: Collaborative robots (cobots) and advanced automation systems can be easily reprogrammed and adapted to new tasks and product configurations. This flexibility enhances the ability to customise production processes.
4. Customer-Driven Production: Industry 4.0 allows for closer collaboration with customers through real-time data exchange. Manufacturers can tailor products and services based on customer input and feedback.

5. Batch and One-Off Production: Industry 4.0 supports batch production and even one-off production runs at competitive costs, making it economically viable to produce highly customised products.

3.2.1.4.1 *Ice-scoring model*

Potential is measured on a scale from 1 to 5 using the ICE (Innovation, Cost, Efficiency) scoring model. The ICE score is based on numerous different parameters relevant to digital design and simulation in the context of Industry 4.0. With its status as a cutting-edge technology that may drastically minimise production costs and lead times, new product development with 3D printing has a high score of 4. Because it improves efficiency, cuts down on mistakes, and speeds up processes generally, order fulfilment automation also scores a 4. This automation technique gets a 4 for innovation because it is a huge step forward in the way that orders are processed and completed thanks to the use of cutting-edge technologies like AI and robotics.

Though the upfront investment in automation systems may be high (3), the long-term cost benefits through lower labor and mistake rates justify the high score. Its efficiency in streamlining the entire order fulfilment cycle and ensuring that orders are processed quickly and accurately earned it a perfect score of 5. With a score of 3, Customer-Driven Production is critical for satisfying specialised requests but isn't always feasible. This method receives a high score of 4 for originality since it marks a radical departure from conventional methods of manufacturing by placing a premium on adaptability to individual client needs. With a score of 5, Batch and One-Off Production demonstrates the flexibility required for the wide variety of production scenarios characteristic of Industry 4.0.

3.3   SWOT ANALYSIS

In the 1960s and 1970s, Albert Humphrey was the first to introduce Strengths, Weaknesses, Opportunities, Threats, otherwise known as SWOT, to businesses. The analysis is broken down into two parts: an internal analysis that takes stock of the company's strengths and weaknesses and an external analysis that surveys the wider marketplace for opportunities and threats. (Figure 3.1)

3.3.1   Strengths and Weaknesses, Opportunities, and Threats of Industry 4.0

Industry 4.0, often known as the Fourth Industrial Revolution, is a strategy developed in 2011 to make Germany's manufacturing sector more globally competitive. Both advantages and difficulties might be expected from the widespread application of Industry 4.0. Industry 4.0's many advantages include cyber-physical systems (CPS) that are faster and more efficient, as well as manufacturing systems that are more adaptable. Weaknesses include worries about loss of employment and the sharing of data and information between sectors. Threats include employment loss and possible cybercrimes, while opportunities include personalisation and speedier service.

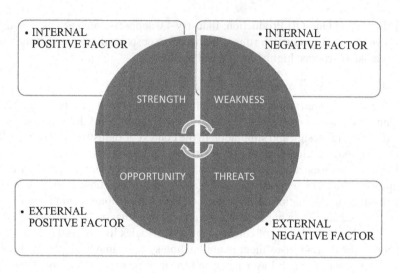

FIGURE 3.1 SWOT analysis matrix.

3.3.1.1 Strengths

It is useful for figuring out how to put an organisation's strengths to work for growth, like its resources, capabilities, and competitive advantages. Organisations may power their expansion by identifying and capitalising on their distinctive strengths.

- It is anticipated that the manufacturing industry's competitive environment will shift as new entrants enter the market and established companies introduce innovative, new products and services.
- Businesses can take stock of what they have and look for unique features they can put to use in the age of Industry 4.0.
- The adoption of Industry 4.0 solutions provides businesses with fresh avenues for expanding their income streams and tapping into previously untapped sources of value.

3.3.1.2 Weaknesses

- Companies may face difficulties due to the complexity and unpredictability of the current market environment.
- In order to survive the shifts introduced by disruptive technologies, it may be necessary for larger, well-established businesses to become nimbler.
- Companies may need to reevaluate their strategy and relationships in light of the emergence of new competitors and the evolution of the value chain.

3.3.1.3 Opportunities

- Opportunities to access new markets and take advantage of untapped value pools are opening up with the advent of Industry 4.0 technology.
- Data and knowledge can be used by businesses to provide more services to customers and generate additional income.

- In order to spur innovation and maintain a competitive edge, businesses can benefit from studying other sectors and adopting their successful business models.

3.3.1.4 Threats

- In the industrial sector, the competitive landscape is expected to become increasingly complicated and uncertain as the number of participants and interfaces grows.
- The value chain could be disrupted by new entrants from industries unrelated to manufacturing, such as telecommunications firms.
- In order to maintain their competitive edge in a dynamic market, businesses must embed agility throughout the entire organisation and prioritise change management.

3.3.2 IMPLICATIONS FOR MANAGEMENT

There are a number of directions that the SWOT analyses conclusions could go for actual industry professionals. Practitioners can better address the risks and shortcomings of Industry 4.0 if they are aware of them. They can use the benefits of Industry 4.0 to protect themselves against its flaws. Effective knowledge management and employee training are essential for a smooth rollout. In sum, the analysis aids professionals in comprehending Industry 4.0 implementation and making sound strategic choices. The results of a SWOT analysis can be put to good use in a variety of ways, some of which are included below.

- A manufacturer may realise that their reputable brand is one of their greatest assets. The corporation might use this strength to its advantage by penetrating untapped markets.
- One of a retailer's flaws can be that it spends too much money on overhead. The business may come up with a plan to save expenses by increasing productivity or negotiating lower rates with vendors.
- The ever-increasing need for cloud-based services presents an opening for some tech firms. The business might take advantage of this opening by creating additional cloud-based offerings or by penetrating new areas with a high demand for such services.
- One of the risks that a financial services firm may face is competition from fintech startups. Investing in cutting-edge technology to remain ahead of the market or creating innovative products and services that fintech firms cannot provide are two options for mitigating this danger.

Ultimately, a SWOT analysis is a useful tool that can aid management in enhancing the organisation's performance and realising its objectives. Management can improve the organisation's plans by taking into account the organisation's SWOT analysis.

3.3.3 COMPARATIVE SWOT ANALYSIS OF INDUSTRY 3.0 VS INDUSTRY 4.0

In Industry 3.0, both mechanical and digital automation are used to streamline manufacturing. Significant efficiency and productivity improvements arose from this [31], but it also had a negative effect on the economy, society, and the environment.

The integration of digital, physical, and biological technology defines Industry 4.0, the fourth wave of the industrial revolution. This is resulting in the creation of novel goods and services, as well as fresh methods of manufacturing and providing them. Complexity, insecurity, the loss of jobs, and societal inequalities are just a few of the problems that Industry 4.0 poses [32,33,34,35].

Some characteristics of Industry 3.0 and Industry 4.0, as well as their advantages, disadvantages, prospects, and dangers, are illustrated below.

3.3.3.1 Strengths

*   In the new Industrial Revolution 3.0, efficiency and productivity have increased dramatically due to mechanical automation and information technology.
    *   Standardised procedures and production techniques.
    *   Reliability in large-scale manufacturing.
    *   Long-standing clientele and market niches.
    *   The cost of technology is reduced initially.
*   Industry 4.0 refers to the use of cyber-physical systems, the Internet of Things [36], and big data analytics to create and distribute novel goods and services.
    *   Extreme malleability and versatility.
    *   Decision-making and data analysis have been improved.
    *   Automation and the Internet of Things to raise productivity.
    *   Production modifications and individualization.

3.3.3.2 Weaknesses

*   High costs of labor and damage to the environment characterise Industry 3.0.
    *   Inability to quickly adjust to new situations.
    *   Wasteful use of available resources.
    *   There has only been a minimal adoption of digital tools.
    *   Threat of interruptions to global supply chains.
*   Complexity and safety concerns in the age of Industry 4.0.
    *   Expenses in starting up are really high.
    *   Workers lack the necessary skills.
    *   Data privacy and security issues.
    *   Difficulties in integrating various existing systems.

3.3.3.3 Opportunities

*   Sector 3.0: Exploring emerging markets and product innovation.
    *   Progressively better processes.
    *   Make the switch to lean production.

TABLE 3.4
Comparative SWOT of Industry 3.0 and 4.0

Industry	Strengths	Weaknesses	Opportunities	Threats
Industry 3.0	Mechanical automation, mass production, efficiency	High labor costs, environmental impact	New markets, product development	Global competition, technological disruption
Industry 4.0	Cyber-physical systems, Internet of Things, big data analytics	Complexity, security risks	Increased productivity, reduced costs	Job displacement, social inequality

- Combining mechanical and electronic processes.
- Emerging market penetration.
- The advent of "Industry 4.0" has led to greater efficiency and lower prices.
  - Predictive upkeep that is much better.
  - Greater insight into the supplier chain.
  - Increased longevity thanks to more efficient utilisation of resources.
  - Development of novel products.

3.3.3.4 Threats
- Global competitiveness and technological upheaval characterise Industry 3.0.
  - Threat from competitors using Industry 4.0 technologies.
  - The lightning-fast decay of technology.
  - Competition for qualified tech workers is high.
  - Environmental compliance regulations present difficult obstacles.
- The Fourth Industrial Revolution and the widening gap between classes.
  - Data breaches and other forms of cyber-danger.
  - Dangers of reliance on modern technology.
  - Existing staff members' resistance to change.
  - Technology becomes obsolete quickly as a result of progress (Table 3.4).

Keep in mind that the aforementioned SWOT analysis matrix is only a high-level outline. Several characteristics, including the size, maturity, and level of technical innovation of the business, will determine the unique strengths, weaknesses, opportunities, and threats that the industry faces [37,38].

3.4 CONCLUSION

Industry 4.0 is widely viewed as a paradigm shift that will fundamentally alter how manufacturing is done in the near future. It includes a plethora of cutting-edge technologies that could greatly benefit businesses in many ways. These include real-time data analysis, higher levels of transparency and visibility, autonomous monitoring, higher levels of productivity, and a more robust competitive edge. Industry 4.0 emphasises horizontal and vertical integration and collaboration.

Financial limitations, lack of technological expertise, organisational limitations, privacy and security issues, and a lack of government policies and support all work against its widespread adoption. As a whole, Industry 4.0 has the potential to revolutionise many different sectors by drastically altering their approaches to product development, management, and customer service. Organisations, industries, and nations all benefit greatly from innovations and technical developments. Industry 4.0 will dramatically alter the products and manufacturing systems in regard to design, processes, operations, and services, but the digital transformation advancements and the expanding interconnectivity will bring new problems to societies. Conventional manufacturing processes are being rethought with the help of many cutting-edge tools and technologies used in "Industry 4.0".

The implementation of Industry 4.0 has the potential to have far-reaching consequences throughout many sectors, including but not limited to the enhancement of production and engineering processes, the enhancement of product and service quality, the optimisation of customer-organization relations, the introduction of new business opportunities and economic benefits, the modification of educational requirements, and the transformation of the current work environment.

Industry 4.0, which involves the digitisation and networking of manufacturing procedures, has promising effects across all three dimensions of sustainability. The KUKA corporation, which operates in fields such as, for example, smart factories, M2M, computing cloud, intelligent robots, e-commerce, etc., applies a number of technologies or applications that assist Industry 4.0 to swiftly differentiate. However, there are also certain obstacles to adopting Industry 4.0 methods. Financial restraints, lack of technical skill of the staff, organisational character, lack of management support and opposition to change, legal concerns, and a lack of policies and backing from the government are just some of the business factors under which these obstacles can be categorised.

Industry 4.0 is sometimes referred to as the "Fourth Industrial Revolution," and this research looks at what it entails in terms of major components, characteristics, implications, drivers, impediments, and implementation issues. Cyber-physical systems, radio frequency identification, the Internet of Things, cloud computing, big data analytics, advanced robotics, and smart factories are all examples of the cutting-edge technologies that make up Industry 4.0.

Industry 4.0 is reshaping several sectors through the incorporation of ICT into business operations, including the automotive, logistics, aerospace, and energy sectors. Companies can benefit greatly from Industry 4.0's capabilities, which include dynamic product creation and development, higher productivity, and increased visibility and autonomy in monitoring and controlling processes. Industry 4.0 is distinguished by its emphasis on horizontal and vertical collaboration and integration. It indicates how the Internet of Things (IoT), artificial intelligence (AI), machine learning (ML), big data analytics (Big Data), blockchain, and robotics may revolutionise current production methods.

Important lessons learned from this study include the following:

- Manufacturing efficiency can be greatly improved when disruptive technologies are integrated in a harmonic manner. Automation, real-time data monitoring, and predictive maintenance all contribute to fewer breakdowns and higher output.
- The chapter highlights the value of data analytics in manufacturing, highlighting the importance of data-driven decision making. Manufacturers may enhance their products, operations, and efficiency by analysing the massive amounts of data they produce.
- Blockchain technology has the potential to transform supply chain management by introducing traceability, authenticity, and trustworthiness to the logistics industry. This safeguards items from manufacture to delivery without a hitch.
- Robotics and automation technologies collaborate with humans to enhance their abilities and free them up to focus on higher-level activities. Overall production quality and efficiency are enhanced by this synergy.
- Disruptive technology deployment isn't without its share of headaches, and this chapter recognises some of them, including high start-up costs, worries about data security, and the need to retrain existing employees. These problems call for strategic responses.
- Evidence from real-world case studies demonstrates how and why disruptive technologies can be useful in a variety of manufacturing settings, from the automobile industry to the electronics industry.
- Potential futures: The authors discussed how the landscape of intelligent manufacturing is changing due to new developments like edge computing and smart factory technology. The importance of considering environmental impact in production is also growing.

3.7 FUTURE OUTLOOK

There is a bright future ahead for the integration of disruptive technology into smart manufacturing. This can anticipate even more cutting-edge and game-changing uses of these technologies in the future as they mature and become more accessible to the general public.

The following are only a few examples of forthcoming trends:

- The application of artificial intelligence and machine learning in manufacturing is already widespread, and it is expected to grow in the next years. These innovations have the potential to streamline operations, enhance quality assurance, and enhance manufacturing.
- IoT and 5G will play a pivotal role in facilitating intelligent manufacturing through the interconnection of machines and the provision of real-time data

collecting and analysis, respectively. Better understanding of operations and more informed decision-making are both possible with this information.

- Robotics and automation will continue to play a significant role in intelligent manufacturing by eliminating human error and risk from previously labor-intensive processes. This will allow people to focus on higher-level, more valuable work.
- Creation of cutting-edge merchandise and services: Intelligent production methods pave the way for ground-breaking innovation in the marketplace. One use of AI and ML is in the creation of tailor-made goods and services for individual consumers.
- New employment and skill sets will be needed as a result of the rise of intelligent manufacturing, which will have a profound impact on the labor market. Skills in artificial intelligence, machine learning, data analytics, and robotics, for instance, will be increasingly in demand.

REFERENCES

1. Lucke, D., C. Constantinescu and E. Westkämper. "Smart factory - a step towards the next generation of manufacturing." *Manufacturing systems and technologies for the new frontier: The 41st CIRP conference on manufacturing systems*, Tokyo, Japan, 2008, pp. 115–118.
2. Wahlster, W. "Semantic technologies for mass customization." In: Wahlster W, Grallert HJ, Wess S, Friedrich H, Widenka T (eds), *Towards the Internet of services*, Springer, Heidelberg, 2014, pp. 3–13.
3. Zakria, Maria, Sumaira Raiz, Muhammad Afzal, Syad Amir Gillani, Abdul Majad. "Knowledge, attitude, practices of water sanitation and hygiene." *Globus An International Journal of Medical Science, Engineering and Technology*, vol. 10, no. 1, 2021, pp. 1–9. doi: 10.46360/globus.met.320211001
4. Vaddadi, Suryaprakasarao, Vishnu Srinivas, Nitish Adi Reddy, Girish H., Rajkiran D. and Anbumani Devipriya. "Factory inventory automation using industry 4.0 technologies." 2022. doi: 10.1109/GlobConET53749.2022.9872416
5. Biral, A., M. Centenaro, A. Zanella, L. Vangelista and M. Zorzi. "The challenges of M2M massive access in wireless cellular networks." *Digital Communications and Networks*, vol. 1, no. 1, 2015. doi: 10.1016/j.dcan.2015.02.001
6. Karnik, N., U. Bora, K. Bhadri, P. Kadambi and P. Dhatrak. "A comprehensive study on current and future trends towards the characteristics and enablers of industry 4.0." *Journal of Industrial Information Integration*, 2021, October, 100294. doi: 10.1016/j.jii.2021.100294
7. Mohammed, A. and L. Wang. "Brainwaves driven human-robot collaborative assembly." *CIRP Annals*, vol. 67, no. 1, 2018. doi: 10.1016/j.cirp.2018.04.048
8. Wang, L., S. Liu, C. Cooper, X.V. Wang and R.X. Gao. "Function block-based simulation of items and actions in the actual world." *Multimedia Tools and Applications*, vol. 67, no. 1, 2013. doi: 10.1007/s11042-012-1013-4.
9. Sunhare, P., R.R. Chowdhary and M.K. Chattopadhyay. "Internet of things and data mining: An application oriented survey." *Journal of King Saud University - Computer and Information Sciences*, In press (1), 2020, pp. 1–22. doi: 10.1016/j.jksuci.2020.07.002
10. Gupta, Tanya and Sameera Khan. "Education and international collaboration in the digital age." *Globus Journal of Progressive Education*, vol. 13, no. 1, 2023, pp. 79–87. doi: 10.46360/globus.edu.220231008

11. Christian, Brecher, Aleksandra Müller, Yannick Dassen and Simon, Storms. "Automation technology as a key component of the Industry 4.0 production development path." *The International Journal of Advanced Manufacturing Technology*, 2021. doi: 10.1007/S00170-021-07246-5

12. Fernandes, Michel M., André Jeferson, Ricardo Bigheti, Pontarolli P. Eduardo and Godoy Paciencia. "Industrial automation as a service: A new application to industry 4.0." *IEEE Latin America Transactions*, 2021. doi: 10.1109/TLA.2021.9480146

13. Tyagi, Amit Kumar, Frederick Fernandez Terrance, Shashvi Mishra and Shabnam Kumari. "Intelligent automation systems at the core of industry 4.0." 2021. doi: 10.1007/978-3-030-71187-0_1

14. Dostatni, Ewa, Jacek Diakun, Damian Grajewski, Radosław Wichniarek and Anna, Karwasz. "Automation of the ecodesign process for industry 4.0." 2018. doi: 10.1007/978-3-319-97490-3_51

15. Büyüközkan, Gülçin, Deniz Uztürk and Öykü Ilıcak. "Digitalization in industry: IoT and industry 4.0." 2020. doi: 10.1201/9780429055621-2

16. Thangamuthu, Mohanraj and R. Jegadeeshwaran. "Introduction to industry 4.0." 2021. doi: 10.1007/978-981-16-3903-6_7

17. Maryam, Abdirad and Krishna K. Krishnan. "Industry 4.0 in logistics and supply chain management: A systematic literature review." *Engineering Management Journal*, 2021. doi: 10.1080/10429247.2020.1783935

18. Kattepur, Ajay. "Workflow composition and analysis in industry 4.0 warehouse automation." 2019. doi: 10.1049/IET-CIM.2019.0017

19. Khari, D., V. Sharma and N. Agarwal. "Effect of pandemic COVID-19 on economic crisis and health issues globally." *Cosmos Journal of Engineering & Technology*, vol. 10, no. 1, 2020, pp. 9–15. doi: 10.4680/cosmos.et

20. Yuan, Xue-Ming. "Impact of industry 4.0 on inventory systems and optimization." 2020. doi: 10.5772/INTECHOPEN.90077

21. Carstensen, J., et al. "Condition monitoring and cloud-based energy analysis for autonomous Mobile manipulation - smart factory concept with LUH bots." *Procedia Technology*, vol. 26, 2016. doi: 10.1016/j.protcy.2016.08.070

22. Xu, T., et. al. "Dynamic identification of the KUKA LBR IIWA robot with retrieval of physical parameters using global optimization." *IEEE Access*, vol. 8, 2020. doi: 10.1109/ACCESS.2020.3000997

23. Ghadge, A., Er M. Kara, H. Moradlou and M. Goswami. "The impact of industry4.0 on supply chains." *Journal of Manufacturing Technology Management*, vol. 31, no. 4, 2020, pp. 669–686. doi: 10.1108/JMTM-10-2019-036

24. Maldonado-Ramirez, A., R. Rios-Cabrera, I. Lopez-Juarez. "A visual path-following learning approach for industrial robots using DRL." *Robotics and Computer-Integrated Manufacturing*, vol. 71, 2021. doi: 10.1016/j. rcim.2021.102130

25. Dwivedi, Archana. "Psychological and technical barrier for teachers to shift face to face to online education during pandemic." *Globus An International Journal of Management & IT A Refereed Research Journal*, vol. 14, no. 2, 2023, pp. 76–80. doi: 10.46360/globus.mgt.120231010

26. Borromeo, Demie S., Noel E. Estrella and Editha R. Caparas. "Impact of educational technology tools on the digital and information literacy skills of selected Dominican Schools in The Philippines." *Cosmos An International Journal of Management*, vol. 12, no. 2, 2023, pp. 1–8. doi: 10.46360/cosmos.mgt.420231001

27. Kumar, Puneet. "Prelude of security dispensation in web technology." *Cosmos Journal of Engineering and Technology*, vol. 10, no. 1, 2020, pp. 5–8. doi: 10.46360/cosmos.et

28. Divyanshu Dixit, Dr. R.S. Parihar. "A study on software library for graph analytics." *Globus an International Journal of Medical Science, Engineering and Technology*, vol. 9, no. 2, 2020, pp. 11–14. doi: 10.46360/globus.met

29. Francisco, Christopher D.C. "Servant leadership: Its influence on collaborative school culture and organizational trust." *Cosmos an International Journal of Art and Higher Education*, vol. 11, no. 1, 2022, pp. 12–26. doi: 10.46360/cosmos.ahe.520221003

30. Al-Daouri, Prof. Zakaria Muttlak and Belal Khalid Atrach. "The impact of strategic intelligence on strategic flexibility in bank Al-Etihad in Jordan." *Globus: An International Journal of Management & IT*, 2020. doi: 10.46360/globus.mgt.120202 007

31. Bebas, Hazel Jill T., et al. "Investigating the self-regulated online learning strategies of bachelor of arts in English language students." *Globus Journal of Progressive Education*, vol. 12, no. 2, 2022, pp. 20–24. doi: 10.46360/globus.edu.220222004

32. Lasi, H., P. Fettke, H.G. Kemper, T. Feld and M. Hoffman. "Industry 4.0." *Business & Information Systems Engineering*, vol. 6, 2014, pp. 239–242. doi: 10.1007/s12599-014-0334-4

33. Alolod, Kim Hazel L., et al. "Effectiveness of online platforms in the education and training of the tourism students." *Cosmos an International Journal of Art & Higher Education*, vol. 11, no. 1, 2022, pp. 37–44. doi: 10.46360/cosmos.ahe.520221005

34. Wuest, T., D. Weimer, C. Irgens and K.D. Thoben. "Machine learning in manufacturing: Advantages, challenges, and applications." *Production & Manufacturing Research*, vol. 4, no. 1, 2016, pp. 23–45.

35. Agarwal, Nidhi. "Role of impact advisory firms in promoting equitable growth in India." *Revista Review Index Journal of Multidisciplinary*, vol. 2, no. 1, 2022. doi: 10.31305/rrijm2022.v02.n01.002

36. Mohanraj, C., et al., "Conspiracy in the stealing of electricity detection through the IOT." *2023 3rd International Conference on Innovative Practices in Technology and Management (ICIPTM)*, Uttar Pradesh, India, 2023, pp. 1–5, doi: 10.1109/ICIPTM57143.2023.10117849

37. Ali, M.A., B. Balamurugan, R.K. Dhanaraj and V. Sharma, "IoT and blockchain based smart agriculture monitoring and intelligence security system." *2022 3rd International Conference on Computation, Automation and Knowledge Management (ICCAKM)*, Dubai, United Arab Emirates, 2022, pp. 1–7, doi: 10.1109/ICCAKM54721.2022.9990243.

38. Raj, A., V. Sharma and A.K. Shanu, "Comparative analysis of security and privacy technique for federated learning in IOT based devices." *2022 3rd International Conference on Computation, Automation and Knowledge Management (ICCAKM)*, Dubai, United Arab Emirates, 2022, pp. 1–5, doi: 10.1109/ICCAKM54721.2022.9990152.

4 Digital Twin Technology Use Cases in Intelligent Manufacturing

Shuchi Sethi, Naeem Ahmad,
Seema Rani, and Misbah Anjum

4.1 INTRODUCTION

The German government has dubbed this new periodas Industry 4.0 or the Fourth Industrial Revolution because it marks the beginning of a new era marked by intelligent machines that will transform our increasingly digital world in 2021. By enabling factories to become production lines, electricity enabled the Second Industrial Revolution. The Third Industrial Revolution in manufacturing began with the convergence of computer and communications technologies, specifically programmable logic controllers (PLCs), which led to automated and improved production processes. The cyber-physical production system (CPPS), a high-information production system in the era of Industry 4.0, allows businesses to place orders by interacting and working in real time with various manufacturers [29–30]. Three primary issues with CPPS automation have been identified by researchers [30]:

1) Modularity and interfaces allow control software to adapt more easily to changing requirements.
2) Inconsistencies within the mechatronic system can be effectively managed and identified through the utilization of semantic technologies, which promote modularity.
3) Urgently gathering and inspecting information as events unfold proves pivotal in rising to meet the pressing difficulty at present. Code paradigms have evolved to incorporate reusable and tested libraries to overcome this challenge, enabling more efficient management.

The application of semantic web technologies has proven many times to be useful in solving the second challenge. Moreover, the emergence of cyber-physical production systems (CPPS) has fulfilled the third challenge. With the help of the IoT and AI, digital models of different cyber-physical systems (CPS) can be obtained in new ways. The IoT has become an important part of CPS as it provides the necessary infrastructure for physical devices to understand, communicate, and interact. Jobs related to the Internet of Things (IoT) include manufacturing, logistics and transportation, energy and electronics, and data and information technology.

CPS is characterized by the wireless connection of smart physical devices that provide lighting, actuation, and control functions. CPS creates a digital twin through continuous physical monitoring and applying computational control to virtual devices. The concept of virtual twins is revolutionizing the capability to replicate real-life situations, including human interactions. This computing power provides the ability to create social cyberspace and expand the boundaries of human imagination. By modeling difficult-to-repeat scenarios, conditions such as the gravity of the moon or the ability of ocean objects to float, can be simulated. In these simulated environments, urban areas can be designed in such a way that survival skills in a variety of environments can be explored and understood. Essentially, the concept of digital twin provides a glimpse into future possibilities [1].

The academic community has proposed various explanations for the use of digital twin counting [2–8]. Figure 4.1 shows the digital model, shadow, and twin.

The dominant point of being a virtual counterpart is to have the following features. It should be a flexible system that can show the synchronized state of the real and virtual representation of the elements at any time:

a. Real-time data
b. Automatic flow of data
c. Bidirectional flow of data

4.2 TRADITIONAL VIEW OF DIGITAL TWIN

Traditionally, a human could not separate the knowledge of physical objects from his or her environment. In the second half of the twentieth century, the collection

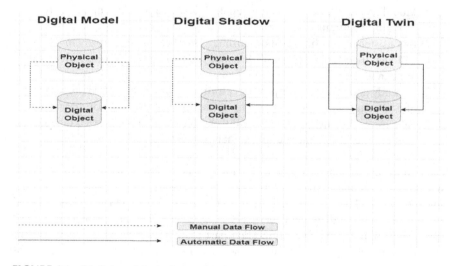

FIGURE 4.1   Digital model, shadow and twin.

of physical information about an object in addition to itself laid the foundation for what was called a digital replica. This process ultimately led to the formation of a virtual twin. As previously mentioned, digitally created virtual replicas mimic the physical properties of real objects while also creating constant forces affecting them, causing their performance to change. A copy or model of the real product can be made before testing so that it is unlikely to be damaged during testing. This process is expensive because it means that the effects of stress are not expected to appear until after actual use. The concept of a product avatar aims to enable consumers to benefit from the services of smart products by creating equivalent products [9–14]. In 2002, the University of Michigan introduced virtual twin technology for product lifecycle management (PLM) to the manufacturing industry. This method involves many digital twin objects such as the digital world and the real world, data transfer between the two, and virtual subspace [2,15].

The term "inventory management" was first used in 2002 to describe the integration of non-standard systems. It was introduced as a mirror space model [16]. Later, the airline industry adopted and expanded this concept. This concept is included in NASA's technology roadmap [17]. The digital twin of the terminal is provided by Virtual Digital Fleet Leader (VDFL), a digital twin technology designed to ensure consistent security throughout the life of the fleet. The technology was designed to overcome the physical limitations of the vehicle. Each level of strategy enables leveraging digital twin capabilities. According to future plans, the Air Force and NASA will need a lighter vehicle that can withstand severe weather conditions and can carry larger objects. These developments require change. Using virtual twin technology, advanced simulation and historical facts are integrated with the vehicle's health control and mirrors to monitor various flight conditions [18]. It has been proposed to use a digital twin to reconstruct the aircraft model with operational deviations and temperatures to account for the current lack of flights [19], and to use digital twins to model the structure to analyze the damage of the confounding pathway [20]. The terms used to describe the concept of digital twin have evolved over time, as seen in Figure 4.2. Digital twins were first defined in the field of aviation, but as they began to be used for production purposes, the word "aircraft" was replaced by "system" in the dictionary. This concept was later applied to many industries, giving rise to new terms such as "digital copy of a physical asset" and "experimental digital twin". Bi-directional data transfer between physical and digital devices is the common point of all these elements and concepts. By Gartner 2017, a three-year study conducted from 2019 to 2022 found the continued importance of digital twin technology [21–24].

Furthermore, it can be argued that the aerospace sector was the first to use digital twin technology, with other manufacturing sectors following suit. We will touch on a few of the many benefits that these machines offer.

First, digital twin technology can replicate an individual's current state of health and has uses in medical and health management. By using cost models, planning patient capacity and expansion, and modeling personnel demands, digital twins have the potential to ameliorate the overall profitability of the healthcare sector. The virtual twin can also detect anomalies, diagnose illnesses, and keep an eye on how

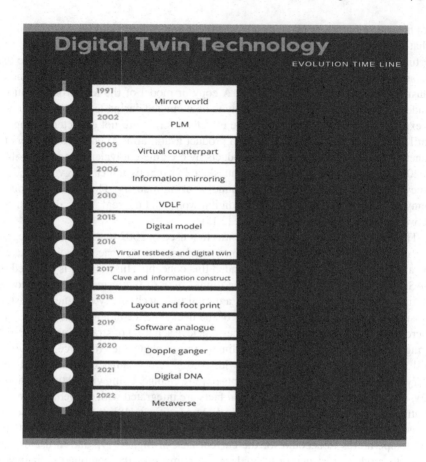

FIGURE 4.2   Time frame of evolution.

well every organ is working. Medical practices would benefit greatly from the use of virtual twins because of their comprehensiveness and accuracy.

Second, dependability and security are increased by using digital twin technology. Like the human body, machines likewise need to have their flaws monitored. The digital twin can identify individual components and identify engine failure. Digital twins are used in manufacturing to predict failure and keep an eye on the work process. When necessary, independent digital twins can be connected to estimate the size of cyberspace. Digital twins can also evaluate a product's condition at any stage of its life.

Third, digital twins provide performance prediction and real-time feedback management. They also encourage the development and use of both novel and traditional methods and goods. Technology optimizes goods, production lines, and processes while ensuring product safety throughout disposal. After a product is out of production, virtual twins can prove useful to build and construct the next one.

Fourth, virtual twins replicate the real world at a certain moment in time using data. This enables the observation and understanding of system performance and predictive maintenance. By representing a system at a detailed level, digital twins aid in product design and manufacturing, allowing for the realization of performance and process improvements. While leveraging current offerings, this facilitates potential avenues for procuring additional income from current solutions.

Last, digital twins play a crucial role in validating product design behaviour and optimizing operational processes in a plant. They offer insightful analysis and prognostications without requiring any sort of negative repercussion. Figure 4.2 shows the digital twin technology's evolution chronology.

4.3 ENABLING TECHNOLOGIES

The DT model combines many technologies, and these technologies combine to become DT technology [27,28]. They fall into one of five categories: IoT, Data Analytics, Large Scale Technologies, Virtual Model technologies and AR/VR technologies. IoT devices that use sensors to supplement input from other sources. It supports the physical parts of DT, including image recognition, particle detection, sensing, reverse engineering, mechanical, aerospace engineering, reliability, and quality engineering. Here the physical strength of the alloy and its ability to withstand pressure and temperature can be defined. In addition to the previously mentioned technology areas, control technology, hydraulics, electric power and gearbox design are also actively working on the latest models of non-power systems. Once data is received, the physical world is recreated with a high degree of accuracy, providing information about features, interactions, and environments. Analysis will be more difficult because the data is heterogeneous in nature and collected from different sources. The data center, where information technology plays a very crucial role in managing response to the environment and adapting to external factors, is a big plus. This field includes technologies for managing and creating information. Bidirectional data transmission is essential for DT technology. The lifetime of DT data consists of the phases of data development, storage, exchange, and interpretation. Large-scale technologies, big data storage frameworks like MySQL and HBase employing Hadoop and NoSQL databases, and Internet of Things technologies like barcodes, QR codes, and RFID, have decreased and added new stress to data transformation, smoothing, cleansing, and compression operations. Anything that is employed in information processing. The evidence suggests that this procedure makes sense. Analytical visualization, predictive analytics, and data mining techniques can all be supported by big data analytics. Effective data communication is facilitated by data visualization. This group also includes instant measurement, machine learning for advanced analysis, word analysis and classification for mining, prediction for bespoke statistics and predictive modeling, and more. Additionally, heterogeneous data cannot be processed or correlated, transmitted, synthesized, filtered, or extracted for important information from several sources without data fusion.

Virtual model technologies, which include geometric, physical, behavioural, and rule models make up another dimension. With the use of computer-aided design software, geometric information—such as size, shape, location, and assembly relationships—can be represented in dimensional and configurational detail. A wide range of software platforms, including AR/VR technologies such as the Unreal Engine, can faithfully depict data about material qualities, density, Poisson's ratio, refractive index, and behavioural modeling derived from physical theories and models. Computational fluid dynamics models, finite element models, robot dynamics models, and the modelling of behaviour using neural networks are among the other technologies needed to depict a virtual model. Technologies for managing constraints and forming relationships depend on the analysis, optimisation, and prediction of object performance. Verification, validation, and certification technologies are used to validate the accuracy of the virtual model, which is another essential component. Service Technology, the fourth dimension, is rooted in applications such as aerospace. Digital Twin (DT) technology offers insights into structural behaviour, lifecycle predictions, and damage estimation within this realm. Within healthcare, the focus lies on meticulously tracking, deducing, and anticipating future wellness states through the delivery of services. To make accurate predictions in these domains, advanced algorithms and AI techniques are necessary for data processing, as well as the decomposition of complex services and intelligent selection of services using visualization technologies. This dimension also encompasses connection and transmission, the fifth dimension, which plays a crucial role in enabling high-precision real-time data transfer to and from the digital world through 5G connectivity. Wireless transmission relies on various supporting technologies, including GPRS, spread spectrum microwave communication, wireless bridge, and satellite communication. These technologies are facilitated by application program interfaces (APIs) that interface between different software systems for seamless data transmission. Like controlling the physical world, the virtual world requires processes regulating behaviour and data transmission. Environmental factors, such as geometric and mechanical parameters specific to marine environments, impact the performance of marine equipment. A deeper understanding of such intricate, interconnected dynamics can be gained by applying simulation technologies like virtual models alongside neural networks merged with augmented reality, virtual reality, and mixed reality.

4.4 INTELLIGENT MANUFACTURING: THE USE OF DIGITAL TWINS

Smart manufacturing, also known as smart manufacturing within the framework of Industry 4.0, uses information technology to plan the production process. Advantages can be achieved by integrating smart devices, new data, data analytics, intelligent tools, and flexible decision-making models across the business. Standards such as ISO 10303 (STEP), which allow data products to be transferred between platforms and levels, facilitate the automation of the production process. Smart manufacturing is characterized by the use of advanced technology, flexibility, and intelligent decision-making based on intelligence. A digital replica can contribute to facilitating the characteristics.

By facilitating the planning, design, and examination of engineering data, as well as permitting investigation into how the functioning model will behave, it aids in analytical work. This model prototype enables data planning at all manufacturing stages. The most recent data from the knowledge base is used in its creation. It makes it possible to forecast how a system will behave; a digital system can be utilized to build a physical system. The two computationally intensive components of digital reproductions are data and models. Data can symbolize diverse bodily qualities, including configuration, warmth levels, and state of being. The model encompasses functions that can recreate different scenarios.

As Figure 4.3 portrays, a manufacturing plant undergoes four distinct stages throughout its lifespan: from its inception during the creation phase to generating products during the production stage, assisting with that during the support phase, and ultimately ending in the disposal stage. In a digital replica, all these phases are interconnected based on usage, and data exchange occurs, or a prototype may be necessary to study the impact of various parameters in the digital environment.

There are two groups of digital transformations: digital twin prototypes, which serve as prototypes, and digital twin instance (DTI), which serve as specific functions. DTP offers all the information needed to produce an online version, including the specifications and expenses related to 3D model creation. But DTI has more data than what's needed for 3D modeling. This data includes component specifications,

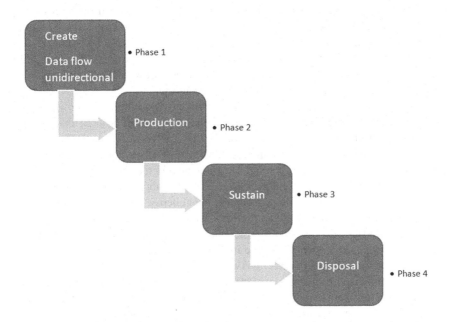

FIGURE 4.3 Lifecycle of manufacturing unit.

past sensor-derived operational or service data, future estimates, and the form and geometrical dimension of physical instances. Instead of being a stand-alone object, the Digital Twin Aggregate (DTA) is an alternative type that is effectively a collection of DTIs. DTA is a stand-alone data structure, but it can query DTIs on-demand or proactively. Ad hoc searches typically include sifting through log data to do computations, like averaging the variation in component failure rates. On the other hand, proactive investigations use real-time data and methods like correlation to anticipate possible component failures.

The use of virtual counterparts can help predict future responses and evaluate physical characteristics. In the past, the creation of physical prototypes accompanied by human descriptions of mental models presented numerous inaccuracies due to humans' limited ability to reproduce mental images, making this task quite challenging. However, the advent of the Digital Twin Prototype has provided a solution. The process operates through several stages, with the initial stage being creation. The physical system is often not readily accessible and only exists within human memory. As explored earlier, this method possesses considerable constraints that diminish its applicability. Complex objects are difficult to describe precisely, requiring significant effort. The four dimensions along which this phase may be carried out include anticipated desirable and undesirable outcomes as well as unforeseen desirable and undesirable developments. The primary objective is to preserve desirable aspects while eliminating undesirable ones.

Using a digital prototype, simulation parameters can be varied across a range, facilitating the investigation of system behaviour in complex scenarios. This allows for avoiding situations that could potentially lead to detrimental outcomes in the event of malfunction or other issues. Following thorough validation, the number of undesirables is minimized, although exploring all possible options is not feasible. Thus, the creation phase enables the mitigation of all undesired aspects. Digital testing of a device's response on a range of individuals with different circumstances becomes feasible in medicine without endangering life.

The subsequent phase, serving as the reverse of the creation phase, is the production phase wherein what was formerly conceived is now brought into fruition. Throughout this phase, the item is fabricated while information streams the opposing way, culminating in the dynamic manifestation of a virtual model symbolizing the physical article. Subsequently, testings are conducted in the sustain phase by switching off parts and states, enabling accurate forecasts about the system's behaviour and the relationship between the actual and virtual items. This phase offers information on expected features in real test versions and real products. Consequently, both anticipated and unexpected undesirables may arise from certain factors that were missed or disregarded throughout the innovation phase. Unlike the earlier phases, data flow is required in both directions during this phase.

Finally, the disposal phase is nevertheless significant while not considered critical. It allows the knowledge base developed throughout the previous three phases to contribute to developing newer product versions. Additionally, valuable lessons regarding proper product disposal and its environmental impact can be learned, thus

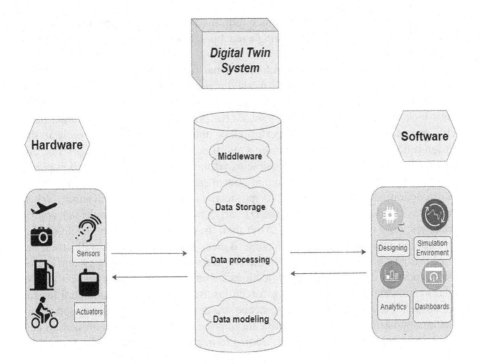

FIGURE 4.4 Digital twin System Model.

aiding in the protection of nature. In Figure 4.4, the digital replica system is shown graphically.

4.5 APPLICATIONS AND USE CASES (UC)

DT technology can be especially valuable in situations where the system is too complex to track down a fault when the system fails [25–26]. Also, in some cases DT can be used in scenarios leading to unpredictable and costly consequences which can sometimes be harmful to human life. By creating digital twins of production machines, DT technology can track performance, assess design functionality, and anticipate potential failures. Healthcare providers can use it to predict illness, predict aviation engine failures, and track working environments, design feasibility, and operational conditions for failure in service invoicing, where data is based on an individual's consumption of the service. A few use scenarios are covered in this section.

UC1: This use case discusses the application of DT in intelligent manufacturing. Complex manufacturing processes, uncertainties, and dynamic disturbances all contribute to performance degradation in rechargeable battery production facilities. Overcoming this architectural resilience is crucial. One thing that DT technology can provide is this: Building resilience requires four steps—diagnosing an occurrence, selecting an action, determining a reaction, and monitoring performance indicators [31–34].

The use of Digital Twin (DT) architecture has fundamentally altered how corporate management is conducted. This method makes use of virtual physical obstacles that facilitate programming, planning, and efficient answers to a variety of tasks. Let's look at the DT design that combines the practical features and product data. This makes it possible to mimic performance and find possible issues early in the design process. Process data including setup time, failure rate, and rework rate are contained in DT. This makes it possible to virtually commission production plans, spot possible bottlenecks, and boost operational effectiveness. This makes it possible to use operational methods in DT for production, supply, and maintenance. When there are deviations from the plan, the DT can identify them and formulate an appropriate response to manage the job. After that, these modifications are made clearer for decision-making and communication. Battery flexibility is used to illustrate the three primary functions of DT. A detailed digital model of the battery system is created, capturing its components, connections, and operational parameters. DT can be used to simulate different operating conditions, helping to identify potential failure points and develop preventive maintenance strategies. Leveraging a DT-based architecture, industries can achieve significant improvements in resource utilization and predict and prevent potential issues, leading to increased resilience and cost savings.

UC2: Whether a product is harmful or not, DT can be quite helpful for testing it in an industry. It is perhaps even more than a sum of the corporeal twin's knowledge and experience. Virtual humans, also known as vhumans, are autonomously capable of difficult tasks and can think and make judgments. For example, following a few meetings, people may collaborate at extraordinarily fast speeds across time and space to reach an unexpected conclusion.

Feelings can be duplicated by recording the condition of the senses and measuring the feelings felt at a given time. It allows one to relive some worthy moments.

It is feasible to create virtual communities that have controlled environments, temperatures, and other elements necessary for human existence. A new metropolis where modes of mobility can be adjusted to investigate what suits people best in particular circumstances and trends that emerge over time. A precise investigation of how the surrounding infrastructure deteriorates over time due to constant humidity and temperature levels is made possible by the duplicated environment's identical lighting, furniture, and other features.

UC3: This case study simulates a manufacturing process. Lifecycle analysis can be performed with precise simulations. Digital twins allow for realistic modeling of the distinguishing characteristics, even while simulation models are highly context specific. Simulating the lifecycle of a system is a four-step process. In Design Simulation, the blueprints for the system are created. Engineers virtually test individual components. In the next step, Virtual Commissioning of the automation system that controls the entire operation is designed and tested in a simulated environment. This helps ensure everything runs smoothly and safely. In the third step, operators are trained using a simulation that mimics real-world scenarios. This helps them learn how to control the system accurately and efficiently. In the fourth step, simulations can be used to monitor sensor functions and optimize production processes.

UC4: Engineers and designers use virtual validation for quality control. In business, virtual validation is becoming more and more prevalent, especially in the automotive industry. The testing and certification of autonomous vehicle functionality is greatly aided by DT. Toyota claims that around 8.8 billion miles of testing would be required to achieve acceptable standards before autonomous vehicles could be introduced. Virtual and semi-virtual testing will be necessary instead of doing this in the actual world due to time, fuel, and space constraints.

UC5: Generative design technologies are shaping the future of DT development and application. Airbus is a major example of the power of DT technology. They use it to explore design options, especially for aircraft cabins. This has led to cases where they have achieved near-optimal configurations–designs that are half the weight of traditional ones. This translates to millions of dollars saved in fuel costs, all while adhering to the strictest safety standards through the development of stronger and more efficient materials.

UC6: Clever manufacturing that is sustainable under this use case, which discusses one of the many ways that DT is applied in the field of automated production and manufacturing. The first is practical gear assembly, which can involve knowledge of unit assembly or presenting insight while the line is being assembled.

With the help of robots and precise tools, the intelligent manufacturing unit houses a variety of smart processing devices. Problem diagnostics can be used to foresee the impacts of overflow shortfalls [35,36]. Punch with force is used by the cold heading machine tool to manufacture a portion out of wire. Using cunning innovations such as new sensors for force-powered solid working and for meeting requirements for fewer fossil fuel byproducts. Among other topics, DT talks about the exploration of alternate fuel for some amazing mining equipment. Second, by using a series of computed heuristic algorithms based on well-organized intelligent manufacturing units, DT in an intelligent production line allows the administration of intelligence and flexibility. When combined, they are speeding up the procedures that result in improvements even on the slowest assembly line.

Services for sustainable intelligent manufacturing [37] include post-purchase assistance, production, and product development. A platform for collaborative designing that links clients and industry professionals is one of the development services. Maintenance and fault diagnosis are included in after-sale services. However, enhanced monitoring and control services are made possible by manufacturing services.

4.6 FUTURE OUTCOMES

The guide ahead is putting DT technology in all aspects of life and making an appropriate guess about how future technology will reshape all industries and all of our lifestyles. An arising area of examination and application is a smart city, which utilizes DT and will include most areas where DT is already in action or can assert its usage in management: traffic management, livestock management, healthcare management, connectivity management, and infrastructure maintenance—technology applications for data analytics, for predictive analytics, data fusion, etc. The scarcity of unified models in the literature demonstrates that the DT technology, as it is, is a perplexing framework requiring further comprehension.

Study and exploration in DT will work with the advancement and variation of Computerized Twin innovation. The consolidation of AI, machine, and deep learning approaches in mounting with DT innovation and has advanced the business, preparing for automation.

4.7 CONCLUSION

This chapter focused on the discussion of DT applications in the manufacturing and aerospace fields. A couple of thoughts under the domain of medical care were noted and any examination connected with progressions in a medical procedures and information combination is being upset because of the delicate data in medical services information about the patients. Addressing the research challenges with standardisation and security of data flow will act as an enabler. Data exchange between IoT devices with more connectivity is emerging as an evident area in Digital Twin systems. Robustness will affirm that the Digital Twin process can bloom when data flows between environments, besides ensuring data exchange is authentic and performs to its calibre.

REFERENCES

1. https://www.rd.ntt/e/dtc/DTC_Whitepaper_en_2_0_0.pdf
2. Grieves, M., & Vickers, J. Digital twin: Mitigating unpredictable, undesirable emergent behavior in complex systems. In *Transdisciplinary perspectives on complex systems.* Springer, Cham, 2017, pp. 85–113.
3. Kritzinger, W., Karner, M., Traar, G., Henjes, J., & Sihn, W. Digital twin in manufacturing: A categorical literature review and classification. *IFAC-PapersOnLine* 2018, 51, 1016–1022.
4. Barricelli, B. R., Casiraghi, E., & Fogli, D. A survey on digital twin: Definitions, characteristics, applications, and design implications. *IEEE Access* 2019, 7, 167653–167671.
5. Liu, M., et al., Review of digital twin about concepts, technologies, and industrial applications. *Journal of Manufacturing Systems* 2020, 58, 346–361.
6. Leiva, C. Demystifying the digital thread and digital twin concepts. *Industry Week* August 2016, 1, 2016.
7. LaGrange, E. Developing a digital twin: The roadmap for oil and gas optimization. In *Proceedings of the SPE Offshore Europe Conference and Exhibition*, Aberdeen, UK, 3–6 September 2019.

8. Singh, M., Fuenmayor, E., Hinchy, E. P., Qiao, Y., Murray, N., & Devine, D. Digital twin: Origin to future. *Applied System Innovation* 2021, 4(2), 36.
9. Kraft, E.M. The air force digital thread/digital twin-life cycle integration and use of computational and experimental knowledge. In *Proceedings of the 54th AIAA Aerospace Sciences Meeting*, San Diego, CA, USA, 4–8 January 2016, p. 0897.
10. Enders, M.R., & Hobach, N. Dimensions of digital twin applications: A literature review. In *Proceedings of the 2019 Americas Conference on Information Systems (AMCIS)*, Cancún, México, 15–17 August 2019.
11. Application of Digital Twin in Industrial Manufacturing. Available online: https://www.futurebridge.com/industry/ perspectives-mobility/application-of-digital-twin-in-industrial-manufacturing/ (accessed on 13 October 2020).
12. Madni, A.M., Madni, C.C., & Lucero, S.D. Leveraging digital twin technology in model-based systems engineering. *Systems* 2019, 7, 7.
13. Maloney, A. The difference between a simulation and a digital twin, 2019. Available online: https://blogs.sw.siemens.com/mindsphere/the-difference-between-a-simulation-and-a-digital-twin
14. Miskinis, C. What does a digital thread mean and how it differs from digital twin? 2018. Available online: https://www.challenge.org/insights/digital-twin-and-digital-thread.
15. Fuller, A., et al. Digital twin: Enabling technologies, challenges and open research. *IEEE Access* 2020, 8, 108952–108971.
16. Grieves, M. Product lifecycle management: the new paradigm for enterprises. *International Journal of Product Development* 2005, 2(1/2), 71–84.
17. https://www.nasa.gov/pdf/501625main_TA12-MSMSM-DRAFT-Nov2010-A.pdf
18. https://arc.aiaa.org/doi/10.2514/6.2012-1818
19. Tuegel, E.J., et al., Reengineering aircraft structural life prediction using a digital twin. *International Journal of Aerospace Engineering* 2011, 2011, Article ID 154798, 14. https://doi.org/10.1155/2011/154798
20. Albert Cerr, J.H, et al. On the effects of modeling as-manufactured geometry: Toward digital twin. *International Journal of Aerospace Engineering* 2014, 2014, Article ID 439278, 10. https://doi.org/10.1155/2014/439278
21. Gartner Identifies the Top 10 Strategic Technology Trends for 2017. https://www.gartner.com/en/newsroom/press-releases/2016-10-18-gartner-identifies-the-top-10-strategic-technology-trends-for-2017
22. Gartner Identifies the Top 10 Strategic Technology Trends for 2018. https://www.gartner.com/en/newsroom/press-releases/2017-10-04-gartner-identifies-the-top-10-strategic-technology-trends-for-2018
23. Gartner Identifies the Top 10 Strategic Technology Trends for 2019. https://www.gartner.com/en/newsroom/press-releases/2018-10-15-gartner-identifies-the-top-10-strategic-technology-trends-for-2019
24. Lorenz, O., Pfeiffer, B., Leingang, C., & Siemens, M.O., 2020. https://www.digitalrefining.com/article/1002413/evolution-of-a-digital-twin-part-1-the-concept#.YtkBROxBxn5
25. Digital Twin Whitepaper, Capgemini, 2019. https://capgemini-engineering.com/as-content/uploads/sites/9/2019/09/digital-twin-pov-whitepaper_v7.pdf
26. Digital Twin Computing, Whitepaper, 2019. https://www.rd.ntt/e/dtc/DTC_Whitepaper_en_2_0_0.pdf
27. Digital refining conference, Feb 2020. https://www.digitalrefining.com/article/1002413/evolution-of-a-digital-twin-part-1-the-concept#.YwV4SHZBy5e

28. Hu, W., et al. Digital twin: A state-of-the-art review of its enabling technologies, appli-
    cations and challenges. *Journal of Intelligent Manufacturing and Special Equipment*
    2021. https://doi.org/10.1108/jimse-12-2020-010
29. Xu, X., et al. Industry 4.0 and industry 5.0—Inception, conception and perception.
    Journal of Manufacturing Systems 2021, 61, 530–535.
30. Vogel-Heuser, B., Industry 4.0–prerequisites and visions. IEEE Transactions on
    Automation Science and Engineering 2016, 13. https://doi.org/10.1109/coase.2007
    .4341667
31. Koulamas, C., & Kalogeras, A. Cyber-physical systems and digital twins in the indus-
    trial internet of things Cyber-physical systems. *Computer* 2018, 51, 95–98. https://doi
    .org/10.1109/MC.2018.2876181
32. He, B., & Bai, K. J. Digital twin-based sustainable intelligent manufacturing: A review.
    Advanced Manufacturing 2021, 9, 1–21. https://doi.org/10.1007/s40436-020-00302-5
33. Zhong, R. Y., & Xu, X. Intelligent manufacturing in the context of industry 4.0: A
    review. *Engineering* 2017, 3(5), 616–630. https://doi.org/10.1016/J.ENG.2017.05.015
34. Park, K.-T., Park, Y.H., Park, M.-W., & Noh, S. D. DT-based CPPS architectural frame-
    work for resilient RBP. *Journal of Computational Design and Engineering* 2023, 10,
    809–829. https://doi.org/10.1093/jcde/qwad024
35. He, B., & Bai, K.J. Digital twin-based sustainable intelligent manufacturing: A review.
    Advances in Manufacturing 2021, 9, 1–21. https://doi.org/10.1007/s40436-020-00302-5
36. Sharma, V., Balusamy, B., Sabharwal, M., & Ouaissa, M. (Eds.). (2023). *Sustainable
    digital technologies: Trends, impacts, and assessments* (1st ed.). CRC Press. https://doi
    .org/10.1201/9781003348313
37. Ahmad, S., Mishra, S., & Sharma, V. Green computing for sustainable future tech-
    nologies and its applications. In Grima, S., Sood, K. and Özen, E. (Ed.) *Contemporary
    studies of risks in emerging technology, Part A (Emerald studies in finance, insurance,
    and risk management)*, Emerald Publishing Limited, Bingley, pp. 2023, pp. 241–256.
    https://doi.org/10.1108/978-1-80455-562-020231016

5 Intelligent Load Migration Using Federated Learning in Intelligent Manufacturing

Ayush Thakur and Nidhi Sindhwani

5.1 INTRODUCTION

Industry 4.0 [1], which integrates cutting-edge technologies like artificial intelligence, big data, cloud computing, and the Internet of Things to improve production processes' efficiency, quality, and sustainability, includes intelligent manufacturing as a fundamental component. But there are other obstacles that intelligent manufacturing must overcome, such as data scarcity, heterogeneity, and privacy. Protecting private or sensitive data from unwanted access or exposure is known as data privacy. The variety and complexity of data sources and formats in intelligent manufacturing systems is referred to as data heterogeneity. The absence of enough or representative data for testing or training machine learning models is referred to as data scarcity.

Federated learning (FL), a promising method that allows collaborative learning among numerous participants without requiring them to share their raw data, is a potential solution to these problems [2]. Using its own data, each party or node may train a local model using FL, a distributed machine learning paradigm. Subsequently, the model parameters can be shared with other nodes or a central server. The parameters can then be combined by the server or nodes to create a global model that incorporates everyone's expertise. Intelligent manufacturing processes including smart process planning, product quality inspection, and tool wear prediction may all benefit from the application of FL.

Intelligent load migration, or the dynamic distribution of computing resources across various devices or network nodes, is one of the possible uses for FL [3]. Because intelligent load migration balances workload and prevents overload or underload conditions, it can enhance system performance and dependability. In a smart factory setting, for instance, certain nodes could be more data- or computation-rich than others, which might lead to production process errors or delays. To improve system stability and efficiency, part of the load can be migrated from the overloaded nodes to the underloaded nodes.

Numerous methods for federated learning-based intelligent load movement in intelligent manufacturing will be discussed in this chapter. Initially, we introduce

DOI: 10.1201/9781032630748-5

the concept and design of FL-based intelligent manufacturing. We next go over our load migration plan based on the FL model parameters and the node statuses. We also discuss the advantages and disadvantages of our technique. Finally, we evaluate our method on a simulated dataset and compare it with a few industry standards.

5.2 FOUNDATIONS OF INTELLIGENT MANUFACTURING

5.2.1 THE EVOLUTION OF MANUFACTURING

Manufacturing is the process of converting raw resources into completed goods that satisfy consumer wants. Throughout history, manufacturing has played a crucial role in human civilisation and has evolved through several stages. We will examine the fundamentals of intelligent manufacturing, which is the manufacturing trend of the present and future that makes use of cutting-edge technologies like robotics, digital twins, artificial intelligence, machine learning, simulation, and self-driving automation to achieve greater sustainability, quality, efficiency, and flexibility [4, 5].

The term "Industrial Revolution" refers to the initial phase of the evolution of manufacturing, which began in Great Britain in the late 1700s and extended to other regions of Europe and North America [6]. The shift from manual labour to mechanised manufacturing utilising steam and water power defined the Industrial Revolution. The mass manufacturing of iron, textiles, and other items was made possible by the development of machinery like the steam engine, power loom, and spinning jenny. With the construction of railroads, canals, roads, and telegraphs, the Industrial Revolution greatly enhanced the infrastructure for communication and transportation. In addition to causing social and environmental issues like urbanisation, pollution, inequality, and exploitation, the Industrial Revolution also boosted the output, income, and population of the industrialised nations.

The Second Industrial Revolution, often referred to as the Technological Revolution, was the second phase of the evolution of manufacturing and lasted from the late nineteenth to the early twentieth century. The invention of new technologies, including those related to electricity, petroleum, steel, chemicals, cars, and aircraft, propelled the Second Industrial Revolution. New modes of entertainment, communication, transportation, and energy production were made possible by these technologies. Along with these new manufacturing techniques, the Second Industrial Revolution also brought in scientific management, the assembly line, and the division of labour. Manufacturing productivity, standardisation, and specialisation all rose as a result of these techniques. Additionally, the Second Industrial Revolution encouraged rivalry and innovation among states and businesses [7].

The Third Industrial Revolution, sometimes referred to as the Digital Revolution, started in the middle of the twentieth century and is still going strong today as the third stage of manufacturing evolution [8]. The development of computers, microelectronics, software, and the internet served as catalysts for the Digital Revolution. The digitisation, automation, and networking of manufacturing systems and processes were made possible by these technologies. New manufacturing paradigms including computer-aided design (CAD), computer-aided manufacturing (CAM), computer-integrated manufacturing (CIM), flexible manufacturing systems (FMS), lean manufacturing, agile manufacturing, and mass customisation were also made possible by the digital revolution [9]. The production process was made more responsive, flexible, efficient, and of higher quality thanks to these paradigms. Along with new opportunities and problems, the digital revolution brought about by globalisation, outsourcing, offshore, e-commerce, cybersecurity, and intellectual property affected manufacturing.

The twenty-first century is witnessing the emergence of the Fourth Industrial Revolution, sometimes referred to as Industry 4.0, which is the fourth stage of manufacturing evolution. Industry 4.0 is defined by the merging of cyber and physical systems through cutting-edge technologies like blockchain, cloud computing, blockchain technology, augmented and virtual reality, robotics, 3D printing, machine learning, Internet of Things, blockchain, artificial intelligence, and big data [5]. Intelligent manufacturing is made possible by these technologies. It is a type of production that can analyse data, adjust to changes, maximise performance, work with people, and run on its own. Comparing intelligent manufacturing to traditional manufacturing, greater levels of efficiency, quality, flexibility, sustainability, innovation, and competitiveness may be attained.

5.2.2 Key Concepts in Intelligent Manufacturing

Intelligent manufacturing is a modern production technique that leverages cutting-edge technologies such as artificial intelligence, machine learning, robotics, cloud computing, and the internet of things to outperform traditional manufacturing in terms of efficiency, quality, flexibility, sustainability, innovation, and competitiveness. Intelligent manufacturing is linked to the notion of Industry 4.0, which is often referred to as the Fourth Industrial Revolution. Cyber and physical system integration is a hallmark of Industry 4.0.

Digitalisation is a fundamental idea in intelligent manufacturing, denoting the process of transforming analogue or physical information into digital or binary form. The process of digitalisation makes it possible to gather, store, process, analyse, and communicate vast volumes of data from many sources and formats. Using software and algorithms, digitalisation also makes it possible to automate and optimise industrial systems and processes. By cutting down on mistakes, waste, downtime, and expenses, digitalisation may enhance manufacturing's quality, efficiency, flexibility, and responsiveness.

Connectivity, or the capacity of machines, devices, processes, and people to communicate and interact with one another through networks, is another essential idea in intelligent manufacturing. Real-time data and information sharing between various entities is made possible by connectivity. Additionally, connectivity makes it possible for manufacturing operations to be coordinated and collaborated upon across several domains and locations. Connectivity can balance workloads, prevent over- or underload scenarios, and enhance decision-making, all of which can lead to increased production performance and dependability.

Intelligence, or the capacity of machines and systems to learn from data, adapt to changes, optimise performance, work with humans, and function autonomously, is a third fundamental idea in intelligent manufacturing [3]. Technologies like artificial intelligence, machine learning, cloud computing, robots, and the Internet of Things make intelligence possible. Many applications in intelligent manufacturing, including load migration, smart process planning, tool wear prediction, and product quality inspection, may be made possible by intelligence. New manufacturing paradigms like digital twins, simulation, augmented reality, virtual reality, 3D printing, and self-driving automation can all be made possible by intelligence. By resolving challenging issues, developing fresh value propositions, and meeting client demands, intelligence may raise the productivity, quality, adaptability, sustainability, innovation, and competitiveness of manufacturing.

5.2.3 Role of Data in Manufacturing Optimisation

Intelligent manufacturing is a contemporary manufacturing strategy that surpasses traditional manufacturing in terms of efficiency, quality, flexibility, sustainability, innovation, and competitiveness. It does this by utilising cutting-edge technologies such as robotics, artificial intelligence, machine learning, the internet of things, cloud computing, and sustainability. The engine of this manufacturing strategy is data. The practice of using data-driven methods and tools to increase the productivity and efficiency of manufacturing systems and processes is known as manufacturing optimisation. Data are necessary for this procedure.

Enhanced visibility and transparency of the production processes is one advantage of leveraging data for manufacturing optimisation. Manufacturers may obtain a better knowledge of the condition and functionality of their systems and processes by gathering and evaluating data from a variety of sources, including equipment, sensors, operators, suppliers, customers, and the environment. In addition, they are able to pinpoint the main causes of issues, the origins of waste and unpredictability, the chances for development, and the most effective methods for reaching success. Additionally, by enabling adaptive adjustments and interventions, aiding decision-making and planning, and offering immediate feedback and warnings, data may assist manufacturers in monitoring and controlling their operations in real-time.

The ability to increase manufacturing output quality and efficiency is another advantage of utilising data for manufacturing optimisation. By using data-driven techniques like advanced analytics, machine learning, simulation, and digital twins, manufacturers can find patterns and relationships between discrete process steps and inputs, optimise the factors that have the biggest impact on yield, predict and prevent failures, reduce defects, improve accuracy, and increase throughput. Data can also enable mass customisation, intelligent process planning, product quality inspection, and consumer feedback analysis, all of which may help manufacturers provide more flexibility and customisation to their goods and services.

Enhancing the sustainability and inventiveness of the production systems is the third advantage of using data for manufacturing optimisation. Manufacturers may lessen their influence on the environment, save resources, abide by laws, and uphold their social duty by utilising data-driven techniques including life-cycle assessment, energy management, waste reduction, and carbon footprint analysis [10]. By facilitating the development of new products, process enhancements, business model transformations, and value proposition formulation, data may assist manufacturers in promoting innovation and competitiveness in their respective markets [11].

However, there are several issues that must be resolved when utilising data for production optimisation. The availability and quality of the data is one of the difficulties. It is imperative for manufacturers to guarantee that the data they gather and utilise is precise, comprehensive, consistent, timely, pertinent, and secure. The obstacles of data integration, data privacy, data scarcity, and heterogeneity must also be surmounted. The use and interpretation of data present another difficulty. To properly analyse and understand the data they have, manufacturers must possess the necessary knowledge, instruments, procedures, and cultural norms. In order to make good use of the data they produce, companies also need to have the appropriate plans, procedures, systems, and governance in place. Realising the usefulness of the data presents a third hurdle. Manufacturers must assess the results of their data-driven projects, including their effect and return on investment. Additionally, they must match the demands of their customers and their company goals with their data-driven objectives [12].

When trying to surmount these obstacles and utilise data for industrial optimisation, firms must implement certain tactics that will facilitate their success on the data-driven path. Creating a precise vision and action plan for their data-driven transformation is one tactic. Manufacturers must define their key performance indicators (KPIs) and targets for data-driven optimisation, assess their current and desired states of data maturity, rank their data-driven initiatives according to viability and value proposition, and schedule the resources and implementation steps.

Creating a solid infrastructure and database is another tactic. Manufacturers must make investments in technologies that will allow them to gather, store, process, analyse, and visualise data from a variety of sources and formats, including sensors, edge devices, cloud platforms, databases, analytics tools, machine learning models, dashboards, and more. To guarantee data quality, availability, and protection, they must also set up data standards, rules, processes, and security measures. Creating a data-driven culture and capacity is the third tactic. Employees at manufacturers must be empowered and trained to use data for creativity, problem-solving, and decision-making. Additionally, they must encourage a culture of cooperation, experimentation, and data-driven learning among all stakeholders, including suppliers, customers, engineers, managers, operators, and so on.

5.3 FEDERATED LEARNING: AN OVERVIEW

5.3.1 BASICS OF FEDERATED LEARNING

Federated learning is a machine learning approach that allows many participants to work together without sharing local data to jointly train a common model. This lowers the transmission and processing costs associated with centralised learning while maintaining the data's confidentiality and privacy. Federated learning has use in several scenarios where data is dispersed among multiple devices or nodes, including edge servers, data centres, cellphones, and Internet of Things devices [13].

Federated learning is predicated on distributed machine learning, which attempts to train a model over several servers or nodes with access to various data subsets. Federated learning, in contrast to distributed machine learning, does not make the assumption that the nodes have comparable compute power and communication bandwidths, nor that the data are independent and identically distributed (IID) [14]. Federated learning, on the other hand, works with non-IID and heterogeneous data that is dispersed throughout erratic and resource-constrained nodes such as laptops, cellphones, and Internet of Things sensors.

Furthermore, federated learning merely requires the nodes to periodically communicate their model parameters—such as a neural network's weights and biases—instead of sending data to a central server or to one another. Federated learning can facilitate collaborative learning among nodes while maintaining the confidentiality and privacy of the data in this way.

A central server and several local nodes make up the two primary parts of the architecture of federated learning. The global model that is shared by every node must be initialised and coordinated by the central server. Using their own data, the local nodes oversee the training of their local models and updating the central server. The general workflow of federated learning is as follows:

- For every training cycle, a subset of nodes is chosen at random by the central server.
- The chosen nodes get the global model parameters from the central server.
- The chosen nodes use a machine learning approach (such as stochastic gradient descent) to train their local models on their own data.
- The central server receives the local model parameters or gradients from the chosen nodes.
- To update the global model, the central server combines the received gradients or parameters using an aggregation algorithm (such as a weighted average).
- Steps 1 through 5 are repeated by the central server until a convergence requirement is satisfied.

5.3.1.1 Federated Averaging

One of the most common algorithms for federated learning is federated averaging, or FedAvg, proposed by Google in 2016 [15]. FedAvg works as follows:

- A subset of nodes that are chosen at random or in accordance with certain criteria get a global model that has been initialised by a central server.
- Every node computes the local model updates after training the model locally on its own data for a predetermined number of epochs or iterations.
- The server receives the local updates from the nodes and weights them according to the quantity of local data samples.

- The aggregated changes are used by the server to update the global model, and the process is repeated until convergence or a predetermined stopping threshold is satisfied.

FedAvg can reduce communication overhead and protect data privacy while achieving performance comparable to or better than centralised learning. But FedAvg also has to deal with a few obstacles, such as:

- The variability of the nodes concerning availability, communication bandwidth, processing power, and data distribution. This might lead to varying convergence rates, accuracy levels, or more or less impact for some nodes than for others on the global model. Several variations of FedAvg have been proposed to address this problem. For example, FedProx penalises large deviations from the global model by introducing a proximity term; FedSGD reduces the number of local training epochs to one; and FedMA uses a multi-attention mechanism to assign different weights to updates from different nodes.
- Being susceptible to rogue or malfunctioning nodes that might provide the server with erroneous or corrupted updates. This might potentially result in adversarial assaults or jeopardise the integrity or quality of the model. Certain techniques have been developed to reduce this risk: FedAvg-Secure uses differential privacy to introduce noise into the local updates or the global model; FedAvg-Robust, which uses robust aggregation methods to detect and filter out outliers; and FedAvg-DP, which uses secure aggregation protocols to prevent eavesdropping or tampering.
- The trade-off between communication cost and accuracy of the model, where more frequent communication may result in higher quality but higher communication costs. This is dependent on a number of variables, including the optimisation goals, data characteristics, model complexity, and network circumstances. Some strategies have been proposed to balance this trade-off: FedAvg-CO reduces the size of the local updates by using compression techniques; FedAvg-AD adapts the communication frequency based on the local gradients through adaptive methods; and FedAvg-MO optimises both communication efficiency and model accuracy through multi-objective optimisation.

5.3.1.2   Privacy-Preserving Techniques

Although federated learning does not require data sharing, it still poses some privacy risks, such as:

- From the model changes or parameters, the server or other nodes could deduce some information about the local data. For instance, they could employ property inference attacks to discover certain dataset attributes (such as illness prevalence or gender distribution) or membership inference attacks to ascertain if a given data point is part of the node's dataset.

- To get access to the global model or updates from other nodes, the nodes could conspire or breach the server. For example, they may employ model extraction attacks to steal the model parameters from the server or other nodes, or they could use model inversion attacks to rebuild certain data samples from the model parameters.
- The nodes might be vulnerable to outside intrusions or eavesdropping, which could reveal their internal data or updates. They could encounter active attacks that alter or introduce harmful data or updates into their network, or passive assaults that watch their network traffic and deduce their data or updates.

Several privacy-preserving methods for federated learning have been proposed to mitigate these problems, including:

- Differential privacy: This mathematical framework adds random noise to the input or output of an algorithm to quantify the privacy loss caused by it. By introducing noise to the local updates or the global model prior to sending or receiving them, federated learning can benefit from differential privacy [16]. This shields the nodes from outside threats and eavesdropping and stops the server or other nodes from deriving the local data from the model changes or parameters. Differential privacy does, however, potentially provide several difficulties, including:
  - The trade-off between utility and privacy, as increasing noise might improve privacy but potentially negatively impact model performance.
  - The challenge of determining the privacy budget, a factor that regulates both the degree of privacy and noise.
  - The difficulty of evaluating the privacy assurances, particularly in environments that are dynamic and diversified.
- Homomorphic encryption: Through the use of cryptographic techniques, encrypted data may be computed without needing to be first decrypted. The global model and local changes can be encrypted via homomorphic encryption, allowing for their safe aggregation or updating without disclosing their contents [17]. This guards against external assaults, eavesdropping, and unauthorised access to or alteration of the data and model by the server or other nodes. Nonetheless, homomorphic encryption has certain difficulties as well, as:
  - The significant cost difference between homomorphic encryption and plain operations, which results in a substantial computational and transmission overhead.
  - Restricted support for intricate processes, such as deep neural networks or non-linear functions.
  - The ability to work with other methods, for example, differential privacy or compression.
- Secure multi-party computation: With the help of this cryptographic technique, several people can collaborate to calculate a function without

disclosing their individual inputs or outputs. Federated learning may be made possible via secure multi-party computing, which enables nodes to train a model together without sharing local data or updates [18]. By doing this, the confidentiality and integrity of the data and model are protected from external threats, eavesdropping, and the server or other nodes. However, there are several difficulties with safe multi-party computation as well, notably:

- The high complexity of communication and calculation stems from the fact that secure multi-party computing protocols need repeated computation and communication between the nodes.
- The protocols' scalability and resilience may be compromised by a high number of nodes, and they may not be able to withstand node failures or dropouts.
- The trade-offs between feasibility and efficiency, as varying levels of security and efficiency may be present in secure multi-party computation protocols, or they may not support all functions or models.

5.3.2 Advantages of Federated Learning

Federated learning is a cutting-edge machine learning method with several benefits, which makes it a desirable option for many applications. Below are listed some of the associated benefits.

5.3.2.1 Decentralised Model Training

Federated learning decentralises the process of training machine learning models, hence bringing about a fundamental transformation. Federated learning enables model training to be dispersed over a large number of edge devices and nodes as an alternative to depending on a single central server or cloud platform [19]. These nodes might be anything from strong data centres to cellphones and Internet of Things gadgets.

This decentralization brings forth several notable benefits:

- Resource Efficiency: Large amounts of equipment and processing power are frequently required for traditional centralised model training. Federated learning lessens the load on cloud or central servers by using the processing capacity of nearby devices. This effective use of resources can result in much cheaper expenses.
- Scalability: Scalability is facilitated by federated learning's decentralised structure. The system can adjust and handle the additional burden as more devices join without requiring significant infrastructure modifications.
- Latency Reduction: Federated learning may cut latency significantly by training models on the edge devices where data is created. Applications that require in-the-moment decision-making, like remote medical diagnostics or driverless cars, especially depend on this [20].

- Resilience: System resilience is increased via decentralisation. The whole model training process may proceed with the remaining devices in the event that one node fails or becomes offline, guaranteeing ongoing development.
- Improved Model Performance: Local data sources frequently have original insights and environmentally relevant trends. Models may access these localised knowledge repositories through federated learning, which might improve the accuracy and performance of the models.

5.3.2.2 Data Privacy

With the advent of data-driven technology, data privacy has become a critical problem. One practical way to protect the parties' security and privacy is through federated learning [21]. Following is a detailed explanation of how federated learning makes this happen:

- Local Data Retention: Data stays on the devices where it is created in federated learning. It is not necessary for parties to divulge their sensitive, unprocessed data to third parties or to each other. By doing this, the danger of data breaches, illegal access, or abuse that comes with centralised data storage is eliminated.
- Privacy-Preserving Techniques: Advanced privacy-preserving methods including federated encryption, safe aggregation, and differential privacy are used in federated learning. These techniques protect user privacy by preventing individual data points from being recreated even throughout the model training phase.
- Regulatory Compliance: Federated learning complies with strict rules and regulations pertaining to data privacy, such as the California Consumer Privacy Act (CCPA) in the US and the General Data privacy Regulation (GDPR) in Europe. By design, it guarantees that the rights of data subjects are protected and reduces the need for significant data exchange.
- Trust and Collaboration: Federated learning encourages cooperation and mutual confidence amongst participants. Together, entities, persons, and organisations may train models without disclosing private information, which makes it a feasible choice for industries like banking or healthcare that have stringent confidentiality regulations.

5.3.3 CHALLENGES AND SOLUTIONS

Federated learning presents a number of issues that need to be resolved in order to be successfully implemented. The two most notable difficulties are listed below.

5.3.3.1 Communication Overhead

Managing the communication overhead between the participating devices, or nodes, and the central server is one of the main issues in federated learning. Network resources may be strained and delays may result from the requirement to communicate model updates, gradients, and other data throughout the training process [22]. Here are a few ways to lessen this difficulty:

- Model Compression: The size of the model updates is decreased by using model compression techniques like quantisation and sparsification. Less bandwidth is needed for smaller updates, which reduces the stress on communication.
- Differential Privacy: The quantity of information disclosed during communication cycles can be restricted at the local level by using differentiated privacy techniques. This permits model training while guaranteeing that the data of each individual participant is kept private.
- Federated Averaging: Devices can update their models locally and submit only aggregated updates to the central server through the use of federated averaging methods. As a result, less data is sent throughout each connection cycle.
- Asynchronous Communication: Communication overhead can be further decreased by introducing asynchronous communication between nodes. Rather than coordinating with every node at once, nodes may independently update the model and send changes whenever they are ready.

5.3.3.2 Non-IID Data

Real-world data is frequently not identically and independently dispersed among devices, or non-IID. This problem stems from data heterogeneity, which is the possibility that several devices will gather data reflecting distinct local characteristics [23]. In order to guarantee that federated learning works, it is imperative that non-IID data be addressed. Here are several methods to overcome this obstacle:

- Data Sampling and Stratification: To make sure that local datasets more accurately reflect the distribution of data generally, devices are capable of doing data stratification and sampling. To do this, representative samples are chosen and data is divided into subgroups based on shared traits.
- Personalisation: Personalisation strategies enable models to leverage the knowledge of the global model while adapting to the specific peculiarities of local data. By doing this, the trade-off between local and global model updates is balanced.
- Transfer Learning: By initialising models with pretrained weights, federated learning can benefit from transfer learning. As it provides a starting point that catches larger patterns in the data, this helps handle non-IID data.
- Data Augmentation: Devices can artificially expand the variety of their local datasets by using techniques for data augmentation. Local data distributions may become more in line with the global distribution as a result.
- Adaptive Aggregation: Differential priorities can be applied to local updates according to the distribution of their data using adaptive aggregation techniques. Updates from devices whose data more closely resembles the world distribution may be aggregated with more weight.

5.4 LOAD MIGRATION IN MANUFACTURING

5.4.1 LOAD BALANCING CONCEPTS

A key strategy in contemporary computing is load balancing, which includes dividing workloads equally among several computing resources, including servers, virtual machines, and containers. Optimising system availability, scalability, and performance is the main objective of load balancing. The network layer, application layer, and database layer are only a few of the tiers of the technological stack where it may be implemented [24,25]. Load balancing is essential for maximising resource use in cloud computing systems and making sure that no resource is overloaded with incoming traffic.

5.4.1.1 Importance of Load Balancing

Cloud-based apps function more efficiently when load balancing is used, and it has several advantages:

- Improved Performance: By distributing workloads uniformly among several resources, load balancing maximises performance. By distributing the load more evenly among the resources, performance bottlenecks are avoided, and the general responsiveness of the system is eventually improved. This is particularly important for applications with fluctuating demand.
- High Availability: The foundation of load balancing is guaranteeing high availability. Load balancers remove single points of failure by dividing up incoming requests among several services. This redundancy ensures that the system can keep running smoothly even in the event of a failure or disruption to one resource. In the event of server outages or unforeseen spikes in traffic, this functionality is essential for preserving service availability.
- Scalability: Scaling resources to meet varying demand is made easier using load balancing. Load balancers have the ability to dynamically distribute traffic to more resources or reduce the utilisation of idle resources when an application's load increases or decreases. Because of their versatility, cloud-based apps can effectively handle fluctuations in demand without the need for human involvement.
- Efficient Resource Utilisation: The foundation of load balancing is efficiency. It keeps any one resource from being overused while others are left underutilised by dividing workloads equally. In addition to optimising system performance, this balanced resource distribution reduces resource waste. In cloud computing, cost optimisation through efficient resource utilisation helps businesses save infrastructure costs.

5.4.1.2 Load Metrics and Metrics Selection

The measurements known as load metrics, which include CPU, memory, response time, throughput, and connection count, are used to gauge the load or performance of a resource. The distribution of the workload among the available resources and the status monitoring of those resources are decided by using load metrics [26].

The selection of load metrics depends on various factors, such as:

- The nature and attributes of the task, including whether it is memory-, CPU-, or network-intensive.
- The nature and attributes of the resources, including whether they are stateful or stateless, static or dynamic, homogenous or heterogeneous.
- The kind of load balancing technique and its properties, including round robin, least connections, least time, hash, or random with two options.
- The system's goals and limitations, include maximising justice, minimising costs, improving dependability, and optimising performance.

For some cases, different load measures may offer distinct benefits and drawbacks. As an illustration:

- One popular load statistic that shows how active a resource is in terms of processing power is CPU utilisation. For CPU-intensive tasks requiring fast computing, it may be helpful. It might not, however, take into account additional elements that impact performance, including memory utilisation or network latency.
- Another popular load statistic that shows how much memory a resource uses is memory utilisation. It could be helpful for tasks requiring a lot of memory and data storage. It might not, however, take into account additional elements that impact performance, including CPU use or network bandwidth [27].
- A load statistic called response time gauges how long it takes a resource to process a request and reply. Workloads requiring high responsiveness and minimal latency may find it helpful. However, other variables like network congestion or client behavior—which are not under the resource's control—may have an impact.
- A load statistic called throughput counts the number of requests a resource can handle in a given amount of time. For tasks requiring great capacity and scalability, it may be helpful. The response time or mistake rate, for example, could not accurately represent the calibre or level of satisfaction with the service.
- A load statistic called "number of connections" indicates how many customers are connecting to a resource at any one moment. Workloads requiring high availability and dependability may find it helpful. However, as various connections could have varying request rates or sizes, it might not accurately reflect the burden or performance of the resource.

As a result, choosing the right load metric is essential to efficient load balancing in cloud computing. A decent load metre should be able to precisely and promptly represent the current condition and capacity of the resource as well as the pertinent characteristics of the workload. Inefficient resource utilisation or inadequate load distribution might result from poor load metrics.

5.4.2 TRADITIONAL LOAD BALANCING APPROACHES

Load balancing is the process of distributing workloads among several computing resources, such as servers, virtual machines, or containers, in order to increase performance, availability, and scalability. Load balancing may be implemented at several stages, such as the database layer, application layer, and network layer. Load balancing is an essential technique in cloud computing for optimising resource utilisation and ensuring that no resource is overwhelmed with traffic [28].

The task of allocating the workload among the available resources may be done using a variety of load balancing techniques. Several prevalent and conventional methods for load balancing are included below.

5.4.2.1 Round Robin

One simple and popular load balancing method is round robin. Its simple yet elegant method of operation involves sending client requests in a sequential, cyclical fashion to each server [29]. Take a look at an example where there are three servers: A, B, and C. Servers A, B, and C receive the first, second, and third inbound requests, respectively. The cycle then continues, sending the fourth request back to server A, and so on.

Advantages of Round Robin:

- Simplicity and Ease of Implementation: The simplicity of the round robin algorithm is one of its main benefits. It is a popular option because it is comparatively simple to comprehend and apply, especially in situations when load balancing has to be put up rapidly.
- Even Load Distribution: The round robin protocol makes sure that client requests are split equally across the servers that are accessible. This fair allocation helps to ensure a balanced use of resources by preventing any one server from being overworked while others are idle [30].
- No Server Left Idle: There is almost never a server that is inactive for long since queries are constantly routed across servers. Because of this, round robin is appropriate in scenarios when it's crucial to maintain server engagement throughout.

Drawbacks of Round Robin:

- Lack of Load Awareness: Round robin does not take each server's capacity or current load into account. It functions under the premise that every server is the same and has the same capacity to process requests. This might not always be the case in practice because servers might differ in terms of CPU, RAM, and other specs. As a result, certain servers can experience overloading more quickly than others, which might cause inconsistent performance or even service outages.
- Request Variability Ignored: Round robin allocates requests to servers based solely on the quantity and kind of incoming requests, treating them

all equally. Not every request is the same in practice; some can require more data or be more sophisticated than others. The latency and performance of the system as a whole may be impacted by the length of time that servers take to answer these requests.

5.4.2.2 Weighted Least Connections

An approach for load balancing that considers both the weight of each server and the quantity of active connections is called weighted least connections. A server's weight is an indicator of its available resources and processing power in relation to other servers. The server can handle more requests the greater the weight [31]. The way the algorithm operates is that client requests are sent to the server that has the fewest active connections compared to its weight. The next request will be forwarded to server C if, for instance, there are three servers, A, B, and C, with weights of 3, 2, and 1, respectively, and server A has six active connections, server B has four active connections, and server C has two active connections. This is because $2/1 < 6/3 < 4/2$.

Advantages of weighted least connections:

- Load and Capacity Awareness: The weighted least connections algorithm's main benefit is its capability to take into account each server's capacity as well as the present load. Administrators can represent a server's memory, processing power, or other pertinent aspects by giving it a weight. As a result, servers with higher capabilities handle a greater portion of the workload, whereas servers with lower capacities handle fewer connections.
- Avoidance of Overloaded or Slow Servers: Sending queries to sluggish or congested services is actively avoided by using weighted least connections. When circumstances change, it automatically reroutes requests to servers that have the capacity to handle them. Users benefit from improved performance and responsiveness as a result.

Drawbacks of weighted least connections:

- Manual Configuration of Weights: The requirement for manual setting is one of weighted least connections' main disadvantages. Based on predetermined standards, administrators have to give each server a weight value. This manual setup might be difficult and less precise in dynamic contexts where circumstances or server capacity can change often.
- Ignorance of Request Variability: Weighted least connections, like other load balancing methods, does not take into account the kind or volume of each request. It is predicated on the idea that every incoming request has comparable processing needs and effects on server load. In actuality, though, certain requests could require more memory or processing power than others. Because of this, certain servers may still suffer from resource fatigue or performance deterioration even with weighted distribution. This is because of the type of requests that these servers get.

5.4.3 LOAD MIGRATION STRATEGIES

The dynamic process of shifting workloads, processes, or tasks from one node (such as a server, virtual machine, or container) to another is known as load migration in distributed systems. It is a key concept in distributed computing [32]. Optimising several elements of system functioning, such as load balancing, performance enhancement, availability assurance, and scalability, is the main goal of load migration. There are two primary categories of load migration: dynamic and static.

Key Objectives of Load Migration:

- Load Balancing: In a distributed system, load migration is frequently used to transfer the computational or resource load evenly among all nodes. This makes sure that no one node gets overworked while other nodes are left unutilised. Reaching load balance helps avoid performance bottlenecks and promotes effective resource usage.
- Performance Optimisation: Load migration aims to maximise the system's overall performance by spreading workloads. In order to improve response times and system responsiveness, tasks may be transferred from nodes that are overloaded to those that have resources available.
- Availability Enhancement: For distributed systems to have high availability, load migration is essential. Load migration minimises service disturbance and maintains continuous operation by rerouting workloads to functional nodes in the event of a node's unavailability due to hardware failure or other problems.
- Scalability: Variations in load can be large for a distributed system in situations that are changing dynamically. Resources may be scaled up or down in response to shifting needs thanks to load migration. Resources can be changed or added as needed to accommodate workload variations.

5.4.3.1 Dynamic Load Migration

In distributed systems, dynamic load migration is a runtime procedure that is determined by the system's present state and load. By spreading jobs or processes among network nodes, it seeks to maximise system availability, performance, and resource use. Depending on the needs and design of the system, a central coordinator, the source node, or the destination node may start a dynamic load migration [33].

5.4.3.1.1 Subtypes of Dynamic Load Migration

Dynamic load migration can be categorised into two primary subtypes:

1. Preemptive Load Migration: Preemptive load migration is a technique where a process or task that is currently executing on the source node is suspended and moved to a destination node where it is restarted [34]. Usually, this kind of migration starts when the source node detects an overload or imbalance. There are several benefits to preemptive load migration.
   - Response Time Reduction: Preemptive load migration can shorten reaction times and boost system throughput by shifting a job from a stressed-out or underperforming node to one that is more capable.

- Resource Utilisation Optimisation: It ensures effective use of the resources that are available by eliminating bottlenecks and balancing resource utilisation across nodes.

Preemptive load transfer does, however, have certain drawbacks, such as higher overhead and complexity. The necessity to preserve and restore the task or process's state after migration gives birth to this complexity, which may have an effect on the system's overall performance.

2. Non-preemptive Load Migration: When a task is in the ready or waiting state, non-preemptive load migration moves it from the source node to the destination node. Non-preemptive migration does not pause ongoing tasks, in contrast to preemptive migration. Rather, its primary focus is on moving non-executing tasks [35]. Some benefits of non-preemptive load transfer are as follows:

   - Reduced Overhead: The migration procedure is made simpler by avoiding the overhead related to pausing and restarting running operations.
   - Resource Efficiency: Because non-preemptive migration might be less disruptive, it is a good choice in situations when it is difficult or impracticable to preserve the task state.

However, the flexibility and efficiency of non-preemptive load movement are limited. It is dependent on the state of the task or process and might not react to imbalances in load as quickly as proactive migration.

5.4.3.2 Load Migration Algorithms

In distributed systems, load migration algorithms are crucial tools for transferring workloads, processes, or tasks across nodes in order to optimise resources, increase system performance, and accomplish load balancing. Usually, these algorithms consist of four essential elements:

1. Transfer Policy: When a node is ready for a task or process transfer, it is determined by the transfer policy component. It entails choosing options according to a range of factors, such as:
   a. Load Threshold: Determining if the load on a node has risen over a certain level, signifying that load redistribution is required.
   b. Load Difference: Assessing how much one node's load differs from the others and determining whether migration is required when imbalances are found.
   c. Load Index: Determining the load index for every node while taking network traffic, memory consumption, and CPU usage into account. These indexes are then used to make migration decisions.
2. Process Selection Policy: The duty of choosing which particular process or task from the source node should be transmitted is the main responsibility of the process selection policy component. This decision's criterion may consist of:

    a. Task/Process Size: When deciding whether tasks or processes are good candidates for migration, taking into account their size or resource needs.

    b. Priority: Giving tasks or processes a number based on their importance and choosing the lower-priority ones to migrate in order to make room for higher-priority work.

    c. Affinity: Determining which operations or processes, because of their compatibility or affinity with particular hardware or software configurations, may be carried out effectively on particular nodes.

    d. Communication Pattern: Examining how activities or processes communicate with one another and choosing those that rely less on inter-node communication.

3. Site Location Policy: The destination node for the task or process migration is chosen by the site location policy component. This choice may be influenced by the following factors:

    a. Availability: Finding nodes with enough resources and availability to support the migration job or process.

    b. Capacity: Determining whether possible destination nodes have the capacity to manage the increased workload.

    c. Proximity: Taking into account nodes' closeness inside the network to reduce latency and maximise efficiency.

    d. Compatibility: Choosing nodes whose hardware and software needs align with the job or process.

4. Information Policy: To assist load migration choices, the information policy component describes how data about the status of the system is gathered and shared across nodes. Possible standards for this policy may be:

    a. Frequency: The frequency at which the system status is tracked and updated to provide precise load balancing choices.

    b. Accuracy: Determining the degree of accuracy and precision of the data gathered, including load measurements, in order to make well-informed migration decisions.

    c. Granularity: Determining the degree of specificity at which data is gathered about a system; this can range from high-level overviews to extremely detailed information.

    d. Scope: Specifying the range of information exchange, including subsets of nodes or the full system.

Some examples of load migration algorithms are:

- Random Algorithm: A straightforward and stochastic method of load migration is the Random Algorithm. It takes a task or process from a source node and moves it to a randomly selected destination node. Although simple, this approach is imprecise and might not be appropriate in situations when more sophisticated load balancing is needed.

- Round-Robin Algorithm: The Round-Robin Algorithm uses a roundrobin system to cyclically choose a task or process from each source node in turn and transfer it to destination nodes. While this approach guarantees some fairness in the distribution of the load, it might not take into consideration differences in task size or system load.
- Threshold Algorithm: Task or process migration from a source node is only initiated by the Threshold Algorithm when its load exceeds a certain threshold. After then, the job is moved to a target node that is under less demand. In situations where loads are within reasonable boundaries, this technique aids in resource conservation and prevents needless migrations.
- Least-Connection Algorithm: Based on how many connections a source node is presently managing, the Least-Connection Algorithm chooses a task or process for migration. It moves the job to a node at the destination that has fewer connections. By dispersing connections, this algorithm seeks to maintain network traffic balance and avoid congestion.
- Genetic Algorithm: An even more advanced method that uses evolutionary processes to discover the best load movement option is the Genetic Algorithm. It functions according to a set of restrictions and an objective function. In order to generate a population of potential solutions, genetic algorithms use processes including crossover, mutation, and selection. This method can handle intricate load-balancing circumstances and is quite flexible.

Since each of these load migration techniques has pros and cons of its own, they may be used in various distributed system use cases. The selection of an algorithm is frequently contingent upon the particular specifications, attributes of the system, and the intended degree of load balancing complexity within the given context.

5.5 INTEGRATION OF FEDERATED LEARNING AND LOAD MIGRATION

Federated learning and load migration are two techniques that may be coupled to increase the efficacy, efficiency, and privacy of distributed machine learning systems. Federated learning enables several nodes to work together to train a shared model without sending their local data, whereas load migration enables moving workloads or processes from one node to another to enhance load balancing [36]. The system benefits from federated learning and load migration in tandem in the below noted ways.

5.5.1 RESOURCE OPTIMISATION

Federated learning plays a role in optimising resource allocation and utilisation in the load migration environment:

- Decreased Communication Overhead: By maintaining data localization on the nodes, federated learning reduces communication overhead. When there is less data flow between nodes, there is less network congestion and load migration latency.
- Effective Model Updates: Federated learning reduces computing cost and complexity in model synchronisation by aggregating model updates or parameters from remote nodes. The smooth integration of model changes into the system is ensured by this simplified method.
- Local Data Adaptation: Federated learning modifies the machine learning model according to the features of individual nodes and the distribution of local data. This enhances the generalisation and accuracy of the model, guaranteeing that it continues to function well even after load migration, which can have a favourable impact on service quality.

5.5.2 REAL-TIME ADAPTATION

The capabilities of federated learning are ideally suited to the dynamic and diverse character of load migration scenarios:

- Iterative Modifications: Federated learning makes incremental and iterative modifications to the machine learning model. Because of this capability, it's well-suited to deal with workload or demand variations that can call for load relocation. Real-time adaptation of the model is possible as the needs of the system change.
- Privacy Protection: Differential privacy and homomorphic encryption are two methods used in federated learning to protect privacy. These methods prohibit unauthorised access or alteration during load transfer, protecting the privacy of the data and the model. This is particularly important when dealing with sensitive data.
- Strong Security Measures: FedAvg-Robust and FedAvg-Secure are two examples of strong and secure aggregation techniques that may be used with federated learning. By using these techniques, the possibility of malicious or malfunctioning nodes jeopardising the machine learning model's quality or integrity during load transfer is reduced.

5.5.3 CHALLENGES AND CONSIDERATIONS

There are additional issues and concerns that must be taken into account when integrating federated learning and load migration in distributed machine learning systemsin the following sections.

5.5.3.1 Data Heterogeneity

When distinct nodes or devices have varied data distributions or properties, this is referred to as data heterogeneity. The following are some ways that data heterogeneity may impact load migration and federated learning:

- Model Performance Degradation: Data heterogeneity may reduce the accuracy or generalization of the model since the global model learned via federated learning may not fit well to each node's data or represent the variety of the data. Data heterogeneity may potentially affect system quality of service or user satisfaction since load migration may not match user preferences or expectations [37].
- Convergence Difficulty: Data heterogeneity may impede federated learning's ability to converge to an ideal solution because local gradients or updates may be biased or inconsistent with the global gradient or model. Because load mobility may result in communication failures or delays, data heterogeneity may also make it more difficult for the nodes and the server to coordinate and synchronise.
- Fairness Issue: Because certain nodes may benefit or have more impact than others in federated learning or load movement, data heterogeneity may lead to injustice or inequality among the nodes or users. Certain nodes could, for instance, acquire more tasks or processes via load migration but have fewer resources or capacity, or they might provide more data or updates to the global model but get less credit or acknowledgment.

To address data heterogeneity, some techniques have been proposed, such as:

- Data Augmentation: This strategy applies techniques like mixup, cutoff, and synthetic data synthesis to boost the variety and similarity of the data among the nodes.
- Data Sharing: Using techniques like random sampling, clustering, or active selection, this method produces a limited subset of data that is globally shared across all the nodes.
- Data Alignment: Using strategies like feature transformation, domain adaptation, or distribution matching, this methodology aligns the data distribution among the nodes.

5.5.3.2 Security and Privacy Concerns

Any threats or assaults that might compromise the confidentiality, availability, or integrity of the data or model in load migration and federated learning are grounds for security and privacy concerns. Concerns about security and privacy can originate from a number of sources, including:

- Nodes that are deemed hazardous may provide malicious or inaccurate updates to the server or other nodes in the context of federated learning, or they may transfer malicious tasks or processes to other nodes through load migration. Malicious nodes might try to steal model parameters, corrupt model performance, infer local data, compromise system integrity, or obtain unfair advantages.
- In load migration, faulty nodes are nodes that inadvertently move problematic jobs or processes to other nodes, or accidentally provide inaccurate or

partial updates to the server or other nodes. Hardware malfunctions, software defects, network mishaps, power outages, or human error can all lead to flawed nodes [37].

- Attacks that come from outside the system are referred to as external attacks. Examples of these include spoofing, replaying, eavesdropping, tampering, denial-of-service, man-in-the-middle, phishing, and ransomware. The data, model, communication route, server, node, user, etc. are all possible targets of external assaults.

Concerns about security and privacy have given rise to several suggested approaches, including:

- Differential Privacy: In federated learning, this method introduces random noise into the global model or local updates to stop information from leaking from the data.
- Homomorphic Encryption: This method secures local updates or the global model in federated learning, enabling their safe aggregation or updating without disclosing their contents.
- Secure Multi-Party Computation: In federated learning, this method enables many nodes to collaboratively calculate a function without disclosing their inputs or outputs.
- Safe Data Aggregation: This method enables nodes to safely combine their local updates in federated learning without sharing them with outside parties or one another.
- Anonymous Communication: This technology uses techniques like onion routing to conceal the location and identity of the nodes in federated learning and load migration.
- Shuffle Model: In federated learning, this method divides and shuffles the local updates at random among several servers in order to thwart collusion or compromise.

5.6 USE CASES AND SUCCESS STORIES

Load migration is a strategy that is used in a variety of scenarios and domains to provide load balancing and optimise resource allocation. The following are a few noteworthy use cases and success stories that emphasise load migration's contribution to increased energy efficiency.

5.6.1 LOAD MIGRATION FOR ENERGY EFFICIENCY

5.6.1.1 Use Case 1: Energy-Efficient Resource Migration in IoT Applications

The load balancing approach presented in this chapter is intended for Internet of Things (IoT) applications. It uses mobile agents to gather data on the IoT system's nodes' resource status. The technique dynamically migrates workloads to the most efficient nodes based on real-time data on resource utilisation and node response times. It hopes to achieve this in order to lower energy usage and enhance the general functionality of IoT applications. Because of the deployment of devices with limited resources, energy efficiency is very important in Internet of Things situations. This is where our technique comes into play.

5.6.1.2 Use Case 2: Dynamic Task Migration for UAV-Enabled Mobile Edge Computing (MEC)

In this study, a task migration technique for mobile edge computing (MEC) systems enabled by unmanned aerial vehicles (UAVs) is introduced. Both load balancing and energy efficiency are taken into account by the algorithm. To find the best work migration plan, it applies an advantage-based value iteration (ABVI) method and a Markov decision process with unknown rewards (MDPUR) model. The main goals are to decrease the overall energy usage of UAVs and efficiently distribute their task, which will improve the energy efficiency of MEC systems. Ensuring timely MEC services and extending UAV flight periods need these kinds of optimisations.

5.6.1.3 Success Story: Load Migration for Renewable Energy Integration

The potential of load migration to encourage the integration of renewable energy sources into power grids and data centre operations is highlighted in this chapter. A key factor in moving electricity consumption from peak to off-peak times is load migration, which makes efficient use of available excess renewable power. It also makes it easier to shift workloads from areas with expensive electricity to areas with less expensive power, which saves money. Load migration helps to minimise greenhouse gas emissions, which in turn promotes environmental sustainability and cleaner energy usage by lowering dependency on fossil fuels and optimising energy utilisation.

These use examples and success stories highlight how load migration may be a tactical tool in a variety of areas, such as IoT, mobile edge computing, and renewable energy integration, to maximise energy efficiency, minimise costs, and minimise environmental effect. The field of load migration techniques is constantly developing and is vital in meeting the increasing need for distributed systems that are sustainable and energyefficient.

5.6.2 Load Migration for Quality Control

In distributed systems, load migration is a useful technique for improving quality assurance and control, especially in areas like manufacturing, testing, and monitoring. Load migration is a deliberate job or process transfer that helps enhance the overall quality, accuracy, and dependability of these systems, which in turn guarantees customer happiness. The following two use cases highlight the function of load movement in quality control.

5.6.2.1 Use Case 1: Distributed Load Management for Manufacturing Systems

Two distributed load management methods are presented in a research study, with an emphasis on large-scale distributed systems such as industrial systems. In the first model, work migration choices are based on a probabilistic normed estimate model, and mobile agents are used to track each node's load status. The second model considers node compatibility, proximity, availability, and capacity while allocating the workload across nodes using fuzzy logic. The goals of both models are to raise the system's overall effectiveness and quality. Their efficient management of load distribution contributes to the reliable and high-quality execution of jobs.

5.6.2.2 Use Case 2: Quality Management Approach for Stock Migration in SAP Systems

A quality management (QM) strategy for stock movement inside systems, applications, and products in data processing (SAP) systems is described in a blog article. It offers a thorough how-to manual for stock migration that preserves the integrity and quality of the data. The method entails verifying stock in the QM view, activating inspection types for goods' receipts from production orders and purchase orders, generating inspection lots for stock posting, and deactivating quality inspection for certain movement kinds. Organisations may avoid mistakes and inconsistencies in their systems by putting these quality control methods in place during stock migration and ensuring that the stock data is dependable and up to the required standards of quality.

These use cases show how load migration may be extremely important for maximising quality control procedures in distributed systems. Through the planned transfer of jobs and the application of quality management techniques, load migration helps ensure that dependable and superior goods or services are delivered, which ultimately benefits consumers and improves the overall operation of the systems involved.

5.7 ARCHITECTURE FOR INTELLIGENT LOAD MIGRATION

The practice of moving jobs or processes from one distributed system node to another while utilising artificial intelligence techniques to maximise load balancing, performance, efficiency, and system privacy is known as intelligent load migration. Applications for intelligent load movement include manufacturing, edge computing, cloud computing, smart grids, and more. An appropriate architecture that facilitates communication, coordination, and computation between the nodes and the server is required to execute intelligent load movement [38–41]. We will talk about several design tenets and elements of an intelligent load migration architecture in the following section.

5.7.1 DESIGN PRINCIPLES

Intelligent load migration architecture should adhere to certain design concepts that represent the needs and goals of the system, noted below.

5.7.1.1 Distributed Architecture

A distributed architecture should include several nodes that are able to interact and work together without the need for a single point of failure or centralised authority. Several advantages of a distributed architecture for intelligent load transfer include:

- Scalability: A distributed architecture may adapt to variations in workload or demand by dynamically adding or deleting nodes.
- Availability: By duplicating or redistributing workloads or processes among several nodes, a distributed architecture can offer high availability and fault tolerance in the event of node failures or network disruptions.
- Privacy: By storing data locally on the nodes and employing encryption or anonymisation techniques to guard against data breaches or leaks, a distributed architecture can maintain the privacy and security of the nodes' data.

5.7.1.2 Scalability and Redundancy

Scalability and redundancy in design refer to the ability of the system to withstand massive volumes of data and requests while maintaining system performance and quality [42]. Many strategies, including the following, can be used to achieve scalability and redundancy:

- Load balancing: To increase system performance and availability, this approach divides the burden among several nodes or servers using techniques such as weighted least connections, least connections, and round robin.
- Caching: To lower system latency and bandwidth, this strategy uses techniques like LRU (least recently used), LFU (least frequently used), or ARC (adaptive replacement cache) to store frequently accessed data or results in memory or disc on the nodes or servers.
- Sharding: To improve system performance and capacity, this approach divides the data into smaller sections, or shards, that are stored on several nodes or servers. It can be applied via hash-based sharding, range-based sharding, or directory-based sharding.
- Replication: To improve the consistency and dependability of the system, this approach makes copies or replicas of the tasks or data on other nodes or servers. It does this by employing techniques like quorum-based replication, peer-to-peer replication, and master-slave replication.

5.7.1.3 Modularity and Reusability

Modular and reusable architecture refers to the arrangement of separate, interchangeable parts that may be merged or utilised in many ways [43]. Intelligent load movement can benefit from various aspects of modularity and reusability, including:

- Flexibility: A modular and reusable design may be tailored to different needs and situations by adding, removing, or altering components as needed to obtain the required functionality and performance.
- Maintainability: A modular and reusable design facilitates testing and development by isolating and fixing errors or issues in individual components without affecting the system as a whole.
- Extensibility: By adding new components or changing old ones without interfering with the system, a modular and reusable design may help with the innovation and integration of new features or technologies.

5.7.1.4 Intelligence and Learning

Artificial intelligence techniques should be employed by the architecture to optimise the load migration process and enable it to learn from data and feedback, so making it intelligent and learning [44]. For intelligent load migration, intelligence and learning can offer a number of advantages, including:

- Efficiency: Through the use of algorithms such as reinforcement learning, genetic algorithms, or deep learning, an intelligent and learning architecture may optimise performance, efficiency, efficacy, and privacy while balancing load.
- Adaptation: An intelligent and learning architecture may adapt to the dynamic and varied environment of load migration by using techniques as online learning, transfer learning, or federated learning to update the model or policy based on the current state and data of the system.
- Personalisation: An intelligent and learning architecture can customise the load migration process for each node or user by using techniques, for instance, as collaborative filtering, content-based filtering, or hybrid filtering to recommend the best tasks or processes to migrate based on the preferences or expectations of the node or user.

5.7.2 Components of the Federated Learning System

A distributed machine learning system known as federated learning allows several nodes to work together to train a common model without sharing local data. The following elements are commonly seen in federated learning systems [45]:

- Nodes: These are the computers or gadgets that take part in federated learning. Examples of these are edge servers, laptops, smartphones, and Internet of Things gadgets. Every node has its own local data and processing capacity. Through a network, it may interact with other nodes or a central server. Nodes may have varying capacity, availability, or data distributions, and they may join or exit the system at any moment.

- Server: This is the main organisation in charge of arranging and enabling federated learning amongst the nodes. The server gathers the local updates from the nodes, updates the global model in accordance with those updates, initialises and maintains the global model, and chooses a subset of nodes to take part in each round of federated learning. Incentives or services as reward distribution, quality control, and privacy protection may also be offered by the server to the nodes.
- Model: This is the machine learning algorithm—a decision tree, linear regression, or deep neural network, for example—that has been trained via federated learning. The weights and biases of a neural network are examples of the parameters that dictate the model's behavior and performance. The model may be trained on a variety of tasks and datasets, including recommendation systems, image classification, and natural language processing.
- Protocol: The protocol is the collection of guidelines and practices that control how the nodes and the server carry out federated learning. In each cycle of federated learning, the protocol specifies how to initialise, synchronise, communicate, aggregate, and update the model. A few more federated learning parameters and requirements, such as the number of nodes, communication frequency, privacy budget, and convergence criterion, may also be specified by the protocol.

5.7.3 DATA PRIVACY AND SECURITY MEASURES

Since federated learning and load migration entail the sharing and processing of sensitive or private data between several parties, data privacy and security are essential components of these technologies. Data privacy and security procedures are intended to both comply with data protection laws and regulations, such as GDPR, and safeguard the data and model from unauthorised access, alteration, or leaking. The following are a few data privacy and security techniques that may be used with load migration and federated learning.

5.7.3.1 Federated Learning and GDPR Compliance

Regulations known as the General Data Protection Regulation (GDPR) set forth guidelines for the collection and use of personal data about individuals inside the European Union (EU) and the European Economic Area (EEA). Individuals have the following rights under the General Data Protection Regulation (GDPR): data portability, consent, access, correction, deletion, limitation, or objection to the processing of their personal data. In addition, the GDPR places a number of requirements on data controllers and processors, including the need to designate data protection officers and carry out data protection impact assessments, along with the responsibility to notify, safeguard, report, and document data processing operations.

By facilitating widespread collaborative learning without revealing or centralizing the original training data, federated learning can assist with GDPR compliance. Federated learning preserves the autonomy and privacy of data subjects by allowing them to retain their data locally on their devices or nodes and to share only the model updates or parameters with other parties. By sending only the modified model parameters or modifications rather than the complete input, federated learning further reduces latency and communication costs.

However, federated learning alone is not sufficient enough for GDPR compliance, as there are still some privacy risks and challenges that need to be addressed, such as:

- From the model changes or parameters, the server or other nodes could deduce some information about the local data. For instance, they could employ property inference attacks to discover certain dataset attributes (such as illness prevalence or gender distribution) or membership inference attacks to ascertain if a given data point is part of the node's dataset.
- To get access to the global model or updates from other nodes, the nodes could conspire or breach the server. For example, they may employ model extraction attacks to steal the model parameters from the server or other nodes, or they could use model inversion attacks to rebuild certain data samples from the model parameters.
- The nodes might be vulnerable to outside intrusions or eavesdropping, which could reveal their internal data or updates. They could encounter active attacks that alter or introduce harmful data or updates into their network, or passive assaults that watch their network traffic and deduce their data or updates.

To address these risks and challenges, some privacy-preserving techniques have been proposed for federated learning, such as:

- Differential privacy: A mathematical framework known as differential privacy measures how much privacy an algorithm loses when random noise is added to the input or output. Federated learning can benefit from differential privacy by introducing noise into the global model or local changes prior to transmission or reception. This shields the nodes from outside threats and eavesdropping, and stops the server or other nodes from deriving the local data from the model changes or parameters.
- Homomorphic encryption: Using a cryptographic technique called homomorphic encryption, encrypted data may be computed without needing to be first decrypted. To enable safe updates or aggregation without disclosing their contents, homomorphic encryption can be utilised to encrypt the local updates or global model. This guards against external assaults and eavesdropping, as well as unauthorised access or alteration by the server or other nodes to the data and model.
- Secure multi-party computation: With the help of this cryptographic technique, several people can collaborate to calculate a function without

disclosing their individual inputs or outputs. Federated learning can be made possible via secure multi-party computing, which enables nodes to cooperatively train a model without sharing local data or changes. By doing this, the confidentiality and integrity of the data and model are protected from external threats, eavesdropping, and the server or other nodes.

5.7.3.2 Federated Learning Frameworks

Software platforms or tools known as federated learning frameworks are used to implement and facilitate federated learning across a range of platforms and devices. Some federated learning functions and features, including data processing, model training, communication optimisation, privacy enhancement, and performance evaluation, are offered by federated learning frameworks. Federated learning frameworks include, for instance:

- TensorFlow Federated: Google created this open-source federated learning framework. TensorFlow, a well-liked machine learning toolkit for creating deep neural networks, serves as its foundation. TensorFlow Federated allows users to express federated computations in Python by means of a high-level Application Programming Interface (API) that supports many types of federated learning scenarios, such as cross-device (where devices link with a central server) or cross-silo (where organizations interact with each other). Additionally, TensorFlow Federated offers a low-level API that lets users modify the federated computations and add their own protocols or algorithms. Differential privacy and safe aggregation are two privacy-preserving strategies that TensorFlow Federated supports [46].
- PySyft: PySyft is an open-source library that supports differential privacy, safe multi-party computing, and federated learning. PyTorch, a well-liked machine learning tool for constructing deep neural networks, serves as its foundation. Using a web-socket protocol, PySyft enables users to carry out federated learning on a variety of devices, including web browsers, cellphones, and edge devices [47]. Additionally, PySyft supports a number of privacy-preserving methods, including differential privacy, safe multi-party computing, and homomorphic encryption. PySyft is a component of the OpenMined project, an open-source community of academics and developers focused on machine learning with privacy preservation.
- FedML: It is an open-source platform designed for research on federated learning. It supports many kinds of devices, such CPUs, GPUs, or mobile devices, as well as different machine learning platforms, like PyTorch, TensorFlow, or scikit-learn. FedML offers a flexible and modular framework that enables users to test and deploy various situations and algorithms for federated learning. Additionally, FedML supports a number of communication optimisation strategies, including gradient tracking, adaptive communication, and compression. The Federated AI Technology Enabler (FATE) project is a federated learning and artificial intelligence research endeavor that is responsible for developing FedML [48].

5.8 IMPLEMENTATION AND USE CASES

5.8.1 Setting Up a Federated Learning Environment

In contemporary distributed machine learning situations, federated learning environments are essential because they provide a strong foundation for cooperative model training across decentralised data sources. We will take you on a thorough tour of the complexities involved in setting up and configuring a federated learning environment in the section that follows. We will cover every step of this complex process, from the initial setup to the management of federated learning systems and the coordination of data sharing protocols among various network nodes.

5.8.1.1 Data Preparation and Partitioning

Data preparation and partitioning are essential elements in the framework of federated learning that require careful consideration and attention to detail. These factors are critical to the efficient functioning of a federated learning environment, and they should be thoroughly investigated.

Primarily, data preparation is the laborious process of selecting and enhancing the unprocessed data obtained from various sources. This process includes several preparatory stages such as data cleansing, normalization, and feature engineering. Different data sources may have different properties, including differences in distribution, format, or quality, which means that every member of the federated learning network will require a different approach to data preparation. In addition to improving data quality, efficient data preparation lowers communication overhead associated with model training.

Concurrently, data partitioning becomes a crucial factor to take into account while using federated learning. A careful distribution of data among the decentralised participants is necessary for proper data partitioning in order to strike a balance between local data representation and security and privacy protection. Data partition designs should take into consideration the unique needs of each participant, bearing in mind the heterogeneity of the data while adhering to security protocols and privacy laws.

Furthermore, the performance and convergence of federated learning models can be significantly impacted by the partitioning strategies chosen, regardless of whether they are based on stratified sampling, random sampling, or other advanced techniques. The necessity for diversity in data representation must be balanced with the need to reduce data exposure or leakage, which creates a difficult situation that requires skillful handling.

5.8.1.2 Federated Learning Framework Selection

Making the correct choice for your federated learning framework is crucial and may have a big impact on your project's outcome. It's critical to consider a number of variables in order to make an informed decision, such as the framework's capabilities, use case applicability, simplicity of integration, scalability, and community support.

5.8.1.2.1　Framework Options

There are several frameworks accessible for exploring the field of federated learning, each with special advantages and skills. Let's summarise a few well-liked federated learning frameworks:

- TensorFlow Federated (TFF): This open-source framework was created by Google and is notable for its strength. With a multitude of tools for model construction, training, and assessment in a federated environment, TFF provides all-encompassing support for federated learning.
- PySyft: A federated learning library with a heavy emphasis on security and privacy, PySyft is built on top of PyTorch. To improve data safety, it integrates cutting-edge methods including homomorphic encryption and secure multi-party computing.
- Federated.ai: Federated.ai is one of the emerging frameworks created especially for edge computing settings and federated learning. These frameworks are designed for situations in which there are many decentralised, edge-based data sources.

5.8.1.2.2　Use Case Considerations

The particular use case at hand should be properly considered while selecting a federated learning system. TensorFlow Federated (TFF), for example, may be a good fit for large-scale, well-established projects that have a need for extensive tooling, whereas PySyft is a great option for apps that handle sensitive data because of its focus on privacy and security.

5.8.1.2.3　Ease of Integration

The simplicity of integrating the chosen framework into your current infrastructure is another important factor to consider. It is advisable to assess compatibility with the programming languages, data storage systems, and machine learning frameworks you currently use. The implementation of federated learning in your company may be accelerated with a smooth integration procedure.

5.8.1.2.4　Scalability

Scalability is a critical issue, especially for large-scale initiatives with increasing participant counts and data quantities. Scalability should be supported by the federated learning framework of choice; to properly meet these changing demands, distributed computing infrastructure may be necessary.

5.8.1.2.5　Community and Support

Last, but not least, the robustness of the framework's user base and the accessibility of copious documentation are crucial components. A lively and helpful community may exchange best practices, provide invaluable support, and help create and improve the framework going forward, which will eventually increase its dependability and usefulness.

5.8.2   Use Case 1: Optimising Production Lines

In the manufacturing industry, optimising production lines is a frequent and crucial operation that may enhance the profitability of the production process, the quality of the final product, and its efficiency. Optimising production lines may be difficult, though, because it requires analysing and predicting the performance and behavior of the production system by applying sophisticated machine learning algorithms to vast volumes of data from a variety of sources, including sensors, equipment, and products. We will demonstrate in this case study how production line optimisation may be achieved in a distributed, privacy-preserving way through the use of federated learning and load migration.

5.8.2.1   Data Collection and Preprocessing

Gathering and preprocessing production system data is the initial stage in optimising production lines. A variety of information types can be included in the data, including machine characteristics, sensor readings, quality labels, product attributes, and failure occurrences. Additionally, the data may be in one of several forms, including textual, categorised, numerical, or temporal. Several strategies, including the following, can be used for the preprocessing and data collection:

- Data acquisition is the method that uses tools like sensors, cameras, or RFID tags to collect and store data from the production system. Depending on the frequency and amount of the data, real-time or batch data capture can be performed.
- Data cleaning is the process of eliminating or fixing mistakes or anomalies from the data that might compromise its correctness or integrity. Techniques including duplicate detection, noise reduction, outlier detection, and missing value imputation can be used to clean up the data.
- Data transformation is the process of changing the data to make it more suited for processing or analysis in a different format. Techniques including discretisation, encoding, normalising, standardisation, and feature extraction can be used to change the data.
- Data integration is the process of combining data from several sources or formats into a single, coherent data collection. Schema matching, entity resolution, record linkage, and data fusion are a few techniques that may be used for data integration.

Every node (a computer or product) involved in load migration and federated learning has the ability to do local preprocessing and data gathering. As a result, each node's raw data may be kept private and secure without having to be shared with a central server or other nodes. Federated learning and load transfer may face difficulties as a result of the data's potential heterogeneity and non-IID among many nodes.

5.8.2.2   Federated Model Training

Using the preprocessed data from the production system, a machine learning model is trained as the second stage of production line optimisation. The model has several applications, including process optimisation, anomaly detection, quality control, and failure prediction [49]. Several techniques, including support vector machine (SVM) [50], random forest (RF), neural network (NN), and long short-term memory (LSTM) [51], can be used to train the model. Federated learning amongst several nodes that take part in load migration and federated learning may be used to carry out the model training cooperatively. In this method, by utilising the dispersed and varied data sources from various nodes, the model performance and generalisation may be enhanced. Nevertheless, this also implies that because various nodes often share model updates or parameters, there may be significant communication and processing costs associated with the model training.

Federated learning works as follows:

- A selection of nodes that are chosen at random or in accordance with certain criteria get a global model that has been initialised by a central server.
- Every node computes the local model updates after training the model locally on its own data for a predetermined number of epochs or iterations.
- The server receives the local updates from the nodes and weights them according to the quantity of local data samples.
- The aggregated changes are used by the server to update the global model, and the process is repeated until convergence or a predetermined stopping threshold is satisfied.

During load transfer and federated learning, a variety of privacy-preserving strategies may be used to safeguard the data and model from unwanted access or alteration. Here are a few instances:

- Differential privacy: To avoid data leaking, this approach adds random noise to the local updates or global model before sending or receiving them.
- Homomorphic encryption: This method encrypts the global model or local updates to enable their safe updating or aggregation without disclosing their contents.
- Secure multi-party computation: This method enables the cooperative computation of a function by several nodes without disclosing their inputs or outputs.

5.9 CHALLENGES AND FUTURE DIRECTIONS

5.9.1 Overcoming Technical Challenges

5.9.1.1 Federated Learning Advancements

Intelligent load transfer has great potential due to federated learning, which enables dispersed learning across several devices or data centres while maintaining data confidentiality and privacy. But in order to reach its full potential, each of the technological difficulties this novel technique presents must be carefully considered. Among these extremely difficult tasks are communication efficiency, heterogeneity, and privacy and security.

Communication Efficiency

Frequent connection between the local devices and the central server is required for federated learning. Its efficiency and scalability may be hampered by this need, which might result in significant bandwidth and latency costs [52]. Researchers have suggested a number of strategies, including aggregation approaches, quantisation, sparsification, and compression of model updates, to lessen these difficulties. These tactics seek to reduce the overhead of communication without compromising the integrity of the educational process.

Heterogeneity

Federated learning utilises a wide range of devices and data sources, all of which have different processing capacities, are available and reliable, and distribute data differently. For federated learning to be implemented successfully, managing this heterogeneity is crucial. To tackle this problem, researchers have developed a number of approaches, such as data sampling methods, tailored models, clustering algorithms, and adjustable learning rates [53]. These methods improve federated learning's flexibility so that it can flourish in a variety of settings.

Privacy and Security

Preserving local data security and privacy is the main goal of federated learning. However, it is still vulnerable to threats like malicious attacks and data leaks. Various security-enhancing strategies have arisen in response to these issues [54]. These include safe multi-party computing, encryption, federated authentication, and differential privacy mechanisms. Federated learning seeks to strengthen its resilience against security flaws and privacy violations by adding these protections.

5.9.1.2 Load Balancing Algorithms

Another essential component of intelligent load migration is load balancing, which optimises system performance and resource consumption by dividing workloads across several servers or devices. However, there are some technological issues that must be resolved before load balancing may be successfully implemented. Several of these multi-faceted problems include complexity, cost considerations, and mitigating points of failure.

Complexity

Load balancing implementation and configuration may be quite difficult tasks, especially when working with large-scale and dynamic systems. For it to function effectively and efficiently, careful planning and design are required. For a load balancing strategy to be deployed successfully, it must strike a balance between system complexity and strategy complexity.

Cost Considerations

When specialist hardware or software solutions are used, load balancing may increase the total cost of the system. System architects must therefore carefully consider trade-offs between cost reduction and performance optimisation. Maintaining load balancing's practicality and economic sustainability requires striking a compromise between these factors.

Mitigating Single Points of Failure
Although load balancing reduces the possibility of a single point of failure inside a system, if it is not done appropriately, it also creates the possibility of becoming a point of failure. Fault tolerance systems and redundancy plans must be combined to strengthen load balancing's dependability and resilience in order to avoid this problem. Sustaining uninterrupted operations requires ensuring system stability in the event of unanticipated breakdowns.

5.10   ETHICAL AND LEGAL IMPLICATIONS

5.10.1   DATA PRIVACY REGULATIONS

Laws governing data privacy control the collection, use, sharing, and storage of personal information by businesses and people. The goals of data privacy laws are to safeguard the interests and rights of data subjects—the people whose data is being processed. The entities that decide the goals and methods of data processing, known as data controllers, and the organisations that process data on their behalf, known as data processors, are also subject to duties and responsibilities under data privacy laws.

Depending on their legal frameworks, cultures, and beliefs, many nations and regions have varying data privacy laws. The following are a few of the most well-known data privacy laws in existence:

- The General Data Protection Regulation (GDPR), which is applicable to all EU member states and any organisation that offers goods or services to EU citizens or monitors their actions, was put into effect by the European Union (EU). It is a comprehensive and uniform data protection law.
- The historic Indian law known as the Digital Personal Data Protection Act (DPDPA) governs how personal data gathered by organisations is managed. Its goal is to safeguard individuals' privacy by giving them control over how their data is processed.
- The GDPR is implemented by the Data Protection Act (DPA) of the United Kingdom (UK), a national legislation that also includes extra requirements for particular industries and circumstances, including national security, journalism, research, and law enforcement.
- The United States (US) state legislation known as the California Consumer Privacy Act (CCPA) gives Californians rights over their personal data, including the ability to view, delete, opt-out, and deal with discrimination.

Federated learning is greatly impacted by data privacy standards since they determine how data may be shared and utilised for cooperative machine learning without jeopardizing data security or privacy. Since federated learning allows for dispersed learning across several devices or data centres without requiring the exchange of raw data, it may offer a way around data privacy laws. Federated learning, however, also has a number of dangers and difficulties with regard to data privacy compliance, including:

- Ensuring data subjects may exercise their rights over their data, including access, correction, deletion, portability, and other rights, and that they have been informed about and given their consent to engage in federated learning.
- Ensuring that data controllers and processors adhere to data minimisation, purpose restriction, accuracy, storage limitation, integrity, confidentiality, etc. and that the processing of personal data for federated learning has a legal basis.
- Ensuring that data transfers between several countries adhere to local rules and regulations in each country and are both safe and legal.
- Ensuring that risk assessments for data privacy and security related to federated learning are carried out in order to identify and reduce those risks.
- Making certain that the proper organisational and technological safeguards are put in place to prevent unauthorised access, disclosure, alteration, or loss of personal data in federated learning.

5.10.2 FAIRNESS AND BIAS CONSIDERATIONS

The ethical ideas of bias and fairness concern how machine learning algorithms treat various groups or individuals according to their traits or qualities. The lack of any unwarranted, institutionalised bias or partiality in machine learning procedures or results is referred to as fairness. Any inaccuracy or divergence in machine learning processes or results stemming from unfair or false presumptions or representations of reality is referred to as bias.

Due to their impact on the performance of federated models for various users or stakeholders in collaborative machine learning, fairness and bias are crucial factors to take into account when implementing federated learning. Because federated learning allows different and heterogeneous data sources to participate in model training without sharing raw data, it may be viewed as a possible chance to increase fairness and minimise bias in machine learning models. However, there are a number of obstacles and problems with fairness and prejudice that federated learning must deal with, including:

- Locating and quantifying bias sources and forms, including communication, aggregation, representation, and selection bias, in federated learning environments.
- Creating and implementing strategies and tactics, such as preprocessing approaches (like reweighing), in-processing techniques (like prejudice elimination), post-processing techniques (like calibration), etc. to reduce bias in federated learning environments.

- Weighing the trade-offs in federated learning contexts between fairness and other goals or restrictions, such as accuracy, efficiency, privacy, security, etc.
- Integrating moral standards and ideals like responsibility, openness, and transparency into the design and management of federated learning.

5.11 FUTURE TRENDS IN INTELLIGENT MANUFACTURING

5.11.1 EDGE COMPUTING AND IoT INTEGRATION

The goal of edge computing and IoT integration is to use edge devices' and sensors' capabilities to process, analyse, and make decisions in real-time for intelligent manufacturing [55]. Intelligent manufacturing may benefit from edge computing and Internet of Things integration in a number of ways, including:

- Better latency and performance: By cutting down on the time and bandwidth needed to transfer data to a centralised cloud or data centre, edge computing and IoT integration can enable real-time analytics and quicker reaction times for vital applications like anomaly detection, quality control, and predictive maintenance.
- Decreased expenses and resource consumption: By optimizing the use of cloud infrastructure and resources, edge computing and IoT integration can lessen the need for high-capacity cloud services and cut down on the price of data transmission and storage.
- Enhanced resilience and dependability: By distributing the computational burden among edge devices, servers, and gateways, edge computing and IoT integration may guarantee ongoing operation in the event of network outages or disturbances. In order to guarantee data availability and integrity, it can also offer redundancy and fault tolerance techniques.
- Enhanced privacy and security: By processing data locally at the edge, edge computing and IoT integration can reduce the risk of breaches and ensure compliance with data protection requirements, protecting sensitive data from unwanted access, disclosure, alteration, or loss.

Some examples of edge computing and IoT integration in intelligent manufacturing are:

- Smart factories: Smart factories monitor and regulate many elements of the manufacturing process, including inventory levels, energy usage, product quality, environmental conditions, and equipment status, using edge computing and IoT integration. Additionally, adaptive production techniques that can adapt to shifting consumer needs, market circumstances, or operational restrictions may be implemented in smart factories through the use of edge computing and IoT connectivity.

- Digital twins: Using edge computing and IoT integration, digital twins are virtualised versions of actual assets or systems that gather data in real time from sensors and devices. By enabling simulation, optimisation, prediction, diagnosis, and insights into the behavior, performance, and health of the physical assets or systems, digital twins can facilitate intelligent manufacturing.
- Collaborative robots: These are machines that can operate side by side with people in sophisticated production settings. Collaborative robots connect with other devices or systems in the network as well as with each other by means of edge computing and IoT integration. Additionally, collaborative robots may learn from human interaction, adjust to changing circumstances, and carry out difficult tasks by utilising edge computing and IoT connectivity.

5.11.2 Federated Learning Innovations

A machine learning approach called federated learning allows groups or several organisations inside one business to work together to jointly train and enhance a shared machine learning model without exchanging raw data. Intelligent manufacturing may benefit greatly from federated learning as it can support data-driven innovation while protecting the confidentiality and privacy of data. Federated learning may help intelligent manufacturing in a number of ways, including:

- Enhanced model quality and accuracy: Federated learning may be used to train more resilient and broadly applicable machine learning models that can manage a variety of situations and tasks in intelligent manufacturing. This is achieved by utilising the varied and heterogeneous data sources from numerous companies or groups.
- Reduced expenses associated with data transmission and storage: Federated learning can improve efficiency and scalability by reducing the volume of data that has to be delivered for model training to a centralised server or cloud, therefore conserving bandwidth and storage resources.
- Enhanced data security and privacy: By storing data locally at the edge and guaranteeing compliance with data protection laws and policies, federated learning can shield sensitive information belonging to several companies or groups from unwanted access, disclosure, alteration, or loss.

Some examples of federated learning innovations in intelligent manufacturing are:

- Federated anomaly detection: This federated learning application looks for anomalous behaviors or occurrences in intelligent industrial systems, such as malfunctioning machinery, flaws, cyberattacks, etc. Through the use of federated anomaly detection, many companies or groups may work together to jointly train and enhance a shared anomaly detection model without exchanging raw data, protecting the confidentiality and privacy of personal information while improving the accuracy and performance of the model.

- Federated reinforcement learning: The goal of federated reinforcement learning is to maximise the control strategies or policies of intelligent industrial systems, including energy consumption, inventory management, production scheduling, and so on. Federated reinforcement learning preserves data confidentiality and privacy while boosting model efficiency and flexibility by allowing disparate companies or groups to work together to jointly learn from and enhance a shared reinforcement learning model without exchanging raw data.

- Federated meta-learning: This is a federated learning application that is made to learn from several related but different IMS jobs, such as defect discovery, quality inspection, and predictive maintenance. Federated meta-learning allows several companies or groups to collaborate to train and improve a shared meta-learning model without sharing raw data, safeguarding data privacy and confidentiality while increasing the model's generality and transferability.

5.12 CONCLUSION

5.12.1 RECAP OF KEY FINDINGS

Using federated learning as a key strategy, we have thoroughly examined the notion and diverse uses of intelligent load movement in the context of intelligent manufacturing in this chapter. The following important conclusions have come out of our conversation:

1. Definition of Intelligent Load Migration: Intelligent load migration is a complex approach that is used to optimise the performance of intelligent manufacturing systems by fine-tuning resource allocation. The dynamic distribution of workloads among a variety of devices or servers enables this. By doing this, the system ensures effective resource usage by adjusting to the shifting conditions and needs.

2. Unveiling Federated Learning: We introduced federated learning, a cutting-edge approach to machine learning. It makes it easier for different companies or groups within one company to work together to train and improve a shared machine learning model. Federated learning is special because it allows for model development without requiring sensitive, raw data to be shared.

3. Advantages of Intelligent Manufacturing: In the field of intelligent manufacturing, there are several benefits of using federated learning for intelligent load movement. These include:
   - Improved Model Quality and Accuracy: Federated learning's collaborative structure combines a variety of data sources to produce machine learning models that are more reliable and accurate.

- Cost-Effectiveness: Intelligent load transfer lowers operating expenses by reducing the need for data transmission and storage.
- Data Security and Privacy: Since federated learning maintains raw data decentralisation, privacy and security issues are mitigated and sensitive industrial data is kept safe.
- System Dependability and Resilience: Load migration's adaptive nature improves system dependability by guaranteeing steady performance even in the face of errors or oscillations.
- Data-Driven Innovation: The utilisation of federated learning cultivates a culture that is focused on data, hence accelerating innovation via insights derived from data.
4. Challenges and Implications: Intelligent load movement using federated learning has several potential benefits, but it also has certain drawbacks that should be carefully considered:
    - Communication Efficiency: High bandwidth and latency expenses may arise from frequent interactions between local devices and central servers. To maximise communication, techniques like quantisation and data compression are crucial.
    - Heterogeneity: Differences in data distributions, computing capacities, and dependability are brought about by the integration of various devices and data sources. To handle this variation, strategies as individualised models and flexible learning rates are essential.
    - Privacy and Security: It is critical to safeguard local data's privacy and security. To solve these issues, encryption, differential privacy, and other security measures are essential.
    - Fairness and Prejudice: When working across various businesses, it's important to take into account the need to ensure fairness in machine learning models and to mitigate prejudice.
    - Ethical and Legal Requirements: To preserve integrity and confidence in the intelligent manufacturing ecosystem, adherence to ethical and legal standards is necessary, particularly with regard to data usage and sharing.

Fundamentally, federated learning-based intelligent load migration denotes a significant paradigm change in the field of intelligent manufacturing. It has the ability to completely transform data cooperation, resource management, and system performance, but it also raises a number of issues that need to be carefully addressed, including ethical issues.

5.12.2 The Synergy of Federated Learning and Load Migration

We find that federated learning and load migration have a remarkable synergy in the field of intelligent manufacturing. These two state-of-the-art methods enhance each other and the system's overall performance and efficiency. Below we examine several such cases where their harmonic interaction is seen:

- Data-Driven Load Migration is Assisted by Federated Learning: Federated learning facilitates load migration by allowing servers or devices to communicate model changes without disclosing private raw data. This methodology significantly reduces transmission overhead while maintaining the highest standards of data security and privacy. This makes the smooth transfer of computing work safer and more effective.
- Migration of Loads Improving Federated Learning: Optimal Distribution of Resources--By wisely distributing resources for model training and inference, load migration enhances federated learning. The system's scalability and computing efficiency are improved by this resource optimization. As a result, the system can quickly adjust to changing needs and complexity, guaranteeing the efficient completion of machine learning assignments.
- Enhancing Load Migration with Federated Learning: Sturdy Model Improvement--Federated learning is the main tool for improving load migration by providing the system with more reliable and accurate models. These models are capable of handling a wide range of situations and assignments in the field of intelligent manufacturing. As a result, the system's overall dependability and performance are significantly increased, enabling it to flourish in contexts that are dynamic and complex.
- Migration of Loads Enhancing Federated Learning: Equitable Distribution of Workloads--Load migration plays a crucial part in enhancing federated learning by dispersing workloads over a heterogeneous landscape of servers or devices. This allocation promotes justice and resilience by guaranteeing the fair use of system resources. As a result, the system is strengthened against inequalities and difficulties arising from device variety.

5.12.3   Implications for the Future of Intelligent Manufacturing

Lastly, we have covered a few of the federated learning and load migration-related future trends and directions in intelligent manufacturing, including federated anomaly detection, federated meta-learning, federated reinforcement learning, edge computing and Internet of Things integration, etc. Federated learning and load migration, which can enable more intelligent, efficient, secure, and innovative manufacturing systems that can handle the growing complexity and diversity of data and applications, are expected to play a significant role in shaping the future of intelligent manufacturing, according to these trends and directions. Consequently, we think that load migration and federated learning are effective methods for raising the bar for intelligent manufacturing.

REFERENCES

1. Lasi, Heiner, et al. "Industry 4.0." *Business & Information Systems Engineering* 6 (2014): 239–242.
2. Li, Li, et al. "A review of applications in federated learning." *Computers & Industrial Engineering* 149 (2020): 106854.

3. Babbar, Himanshi, Shalli Rani, and Salman A. AlQahtani. "Intelligent edge load migration in SDN-IIoT for smart healthcare." *IEEE Transactions on Industrial Informatics* 18.11 (2022): 8058–8064.
4. Carlsson, Bo. "The evolution of manufacturing technology and its impact on industrial structure: an international study." *Small Business Economics* 1 (1989): 21–37.
5. Doll, William J., and Mark A. Vonderembse. "The evolution of manufacturing systems: towards the post-industrial enterprise." *Omega* 19.5 (1991): 401–411.
6. Ashton, Thomas Southcliffe. *The industrial revolution 1760–1830.* Oxford University Press, 1997.
7. De Vries, Jan. "The industrial revolution and the industrious revolution." *The Journal of Economic History* 54.2 (1994): 249–270.
8. Jänicke, Martin, and Klaus Jacob. "A third industrial revolution?" *Long-term governance for social-ecological change.* Routledge, 2013: 47–70.
9. Radhakrishnan, Pezhingattil, S. Subramanyan, and V. Raju. *Cad/Cam/Cim.* New Age International, 2008.
10. Aguado, Sergio, Roberto Alvarez, and Rosario Domingo. "Model of efficient and sustainable improvements in a lean production system through processes of environmental innovation." *Journal of Cleaner Production* 47 (2013): 141–148.
11. Majeed, Arfan, Jingxiang Lv, and Tao Peng. "A framework for big data driven process analysis and optimization for additive manufacturing." *Rapid Prototyping Journal* 25.2 (2018): 308–321.
12. Jihong, Z. H. U., et al. "A review of topology optimization for additive manufacturing: Status and challenges." *Chinese Journal of Aeronautics* 34.1 (2021): 91–110.
13. Bonawitz, Keith, et al. "Towards federated learning at scale: System design." *Proceedings of Machine Learning and Systems* 1 (2019): 374–388.
14. Zhang, Ticao, and Shiwen Mao. "An introduction to the federated learning standard." *GetMobile: Mobile Computing and Communications* 25.3 (2022): 18–22.
15. Li, Xiang, et al. "On the convergence of fedavg on non-iid data." *arXiv preprint arXiv:1907.02189* (2019).
16. Wei, Kang, et al. "Federated learning with differential privacy: Algorithms and performance analysis." *IEEE Transactions on Information Forensics and Security* 15 (2020): 3454–3469.
17. Zhang, Chengliang, et al. "{BatchCrypt}: Efficient homomorphic encryption for {Cross-Silo} federated learning." *2020 USENIX annual technical conference (USENIX ATC 20).* 2020.
18. Byrd, David, and Antigoni Polychroniadou. "Differentially private secure multi-party computation for federated learning in financial applications." *Proceedings of the First ACM International Conference on AI in Finance.* 2020.
19. Zhang, Tuo, et al. "Federated learning for the internet of things: Applications, challenges, and opportunities." *IEEE Internet of Things Magazine* 5.1 (2022): 24–29.
20. Tang, Fengxiao, et al. "Federated learning for intelligent transmission with space-air-ground integrated network (SAGIN) toward 6G." *IEEE Network.* 2022.
21. Shen, Sheng, et al. "From distributed machine learning to federated learning: In the view of data privacy and security." *Concurrency and Computation: Practice and Experience* 34.16 (2022): e6002.
22. Mammen, Priyanka Mary. "Federated learning: Opportunities and challenges." *arXiv preprint arXiv:2101.05428.* 2021.
23. Ma, Xiaodong, et al. "A state-of-the-art survey on solving non-IID data in federated learning." *Future Generation Computer Systems* 135 (2022): 244–258.
24. Li, Di, et al. "A big data enabled load-balancing control for smart manufacturing of Industry 4.0." *Cluster Computing* 20 (2017): 1855–1864.

25. Jafarnejad Ghomi, Einollah, Amir Masoud Rahmani, and Nooruldeen Nasih Qader. "Service load balancing, task scheduling and transportation optimisation in cloud manufacturing by applying queuing system." *Enterprise Information Systems* 13.6 (2019): 865–894.

26. Galbraith, Lissa, and Timothy J. Greene. "Manufacturing system performance sensitivity to selection of product design metrics." *Journal of Manufacturing Systems* 14.2 (1995): 71–79.

27. Naidu, Sasikumar, Rapinder Sawhney, and Xueping Li. "A methodology for evaluation and selection of nanoparticle manufacturing processes based on sustainability metrics." *Environmental Science & Technology* 42.17 (2008): 6697–6702.

28. Collotta, Mario, et al. "Dynamic load balancing techniques for flexible wireless industrial networks." *IECON 2010-36th Annual Conference on IEEE Industrial Electronics Society. IEEE*, 2010.

29. Moylan, Shawn, Christopher U. Brown, and John Slotwinski. "Recommended protocol for round-robin studies in additive manufacturing." *Journal of Testing and Evaluation* 44.2 (2016): 1009–1018.

30. Du Plessis, Anton, et al. "Laboratory X-ray tomography for metal additive manufacturing: Round robin test." *Additive Manufacturing* 30 (2019): 100837.

31. Gill, Amarjit, Nahum Biger, and Neil Mathur. "The relationship between working capital management and profitability: Evidence from the United States." *Business and Economics Journal* 10.1 (2010): 1–9.

32. Wang, Ke. "Migration strategy of cloud collaborative computing for delay-sensitive industrial IoT applications in the context of intelligent manufacturing." *Computer Communications* 150 (2020): 413–420.

33. Jain, Saurabh, and Varsha Sharma. "Enhanced load balancing approach to optimize the performance of the cloud service using virtual machine migration." *International Journal of Engineering and Manufacturing* 7.1 (2017): 41–48.

34. Kim, Sungwook, and Pramod K. Varshney. "Adaptive load balancing with preemption for multimedia cellular networks." *2003 IEEE Wireless Communications and Networking, 2003. WCNC 2003.* Vol. 3. IEEE, 2003.

35. Belgaum, Mohammad Riyaz, et al. "Load balancing with preemptive and non-preemptive task scheduling in cloud computing." *2017 IEEE 3rd International Conference on Engineering Technologies and Social Sciences (ICETSS), IEEE*, 2017.

36. Yuan, Peiyan, et al. "Accuracy rate maximization in edge federated learning with delay and energy constraints." *IEEE Systems Journal* (2022). https://doi.org/10.1109/jsyst .2022.3203727

37. Guendouzi, Badra Souhila, et al. "A systematic review of federated learning: Challenges, aggregation methods, and development tools." *Journal of Network and Computer Applications* (2023): 103714. https://doi.org/10.1016/j.jnca.2023.103714

38. Kumar, Sunil, et al. "Energy efficient resource migration based load balance mechanism for high traffic applications IoT." *Wireless Personal Communications* 127.1 (2022): 385–403.

39. Ouyang, Wu, et al. "Dynamic task migration combining energy efficiency and load balancing optimization in three-tier UAV-enabled mobile edge computing system." *Electronics* 10.2 (2021): 190.

40. Thilagavathi, N., et al. "Energy efficient load balancing in cloud data center using clustering technique." *International Journal of Intelligent Information Technologies (IJIIT)* 15.1 (2019): 84–100.

41. Jin, Qimin, et al. "A distributed fog computing architecture supporting multiple migrating mode." *2018 5th IEEE International Conference on Cyber Security and Cloud Computing (CSCloud)/2018 4th IEEE International Conference on Edge Computing and Scalable Cloud (EdgeCom).* IEEE, 2018.

42. Baptista, Tibério, Luís Bastião Silva, and Carlos Costa. "Highly scalable medical imaging repository based on Kubernetes." *2021 IEEE International Conference on Bioinformatics and Biomedicine (BIBM)*. IEEE, 2021.
43. Dai, William Wenbin, and Valeriy Vyatkin. "A case study on migration from IEC 61131 PLC to IEC 61499 function block control." *2009 7th IEEE International Conference on Industrial Informatics*. IEEE, 2009.
44. Kumar, Yogesh, Surabhi Kaul, and Yu-Chen Hu. "Machine learning for energy-resource allocation, workflow scheduling and live migration in cloud computing: State-of-the-art survey." *Sustainable Computing: Informatics and Systems* 36 (2022): 100780.
45. Lo, Sin Kit, et al. "FLRA: A reference architecture for federated learning systems." *European Conference on Software Architecture*. Cham: Springer International Publishing, 2021.
46. Li, Qinbin, et al. "A survey on federated learning systems: Vision, hype and reality for data privacy and protection." *IEEE Transactions on Knowledge and Data Engineering* 35.4 (2021): 3347–3366.
47. Ziller, Alexander, et al. "Pysyft: A library for easy federated learning." *Federated Learning Systems: Towards Next-Generation AI* (2021): 111–139. https://doi.org/10.1007/978-3-030-70604-3_5
48. He, Chaoyang, et al. "Fedml: A research library and benchmark for federated machine learning." *arXiv preprint arXiv:2007.13518* (2020).
49. Zhang, Chen, et al. "A survey on federated learning." *Knowledge-Based Systems* 216 (2021): 106775.
50. Navia-Vázquez, A., Roberto Díaz-Morales, and M. Fernández-Díaz. "Budget distributed support vector machine for non-id federated learning scenarios." *ACM Transactions on Intelligent Systems and Technology (TIST)* 13.6 (2022): 1–25.
51. Zhao, Ruijie, et al. "Intelligent intrusion detection based on federated learning aided long short-term memory." *Physical Communication* 42 (2020): 101157.
52. Shahid, Osama, et al. "Communication efficiency in federated learning: Achievements and challenges." *arXiv preprint arXiv:2107.10996* (2021).
53. Ye, Mang, et al. "Heterogeneous federated learning: State-of-the-art and research challenges." *ACM Computing Surveys* 56.3 (2023): 1–44.
54. Ma, Chuan, et al. "On safeguarding privacy and security in the framework of federated learning." *IEEE Network* 34.4 (2020): 242–248.
55. Zhu, Wuji, Mohammad Goudarzi, and Rajkumar Buyya. "FLight: A lightweight federated learning framework in edge and fog computing." *Software: Practice and Experience* (2023). https://doi.org/10.1002/spe.3300

6 Challenges and Issues in Securing Intelligent Manufacturing Systems

Gautam Samblani and Devershi Pallavi Bhatt

6.1 INTRODUCTION

In a plant where machines can autonomously think and collaborate consistently to deliver high-quality items at negligible costs, we see the reality of intelligent manufacturing. These systems speak to the cutting edge of generation frameworks, revolutionising the fabricating industry. Made persistently over decades, intelligent manufacturing is incrementally picking up basic balance. The future anticipates a move towards smart generation lines that leverage cyber-physical systems (CPS), also known as cyber physical era systems. Intelligent Manufacturing Systems (IMS) handle the control of Artificial Intelligence (AI), the Internet of Things (IoT), and large-scale data analytics to optimise capability, effectiveness, quality, and sustainability. With smart manufacturing, sensors play a noteworthy role in gathering data from distinctive stages of the manufacturing handle, checking suppliers, equip, shapes, things, and clients. This data amalgamation offers a comprehensive and correct diagram of the entire operation, locks in makers to immediately recognise and redress issues, updates capability, and makes taught commerce choices. In essence, intelligent manufacturing leverages data to improve the effectiveness and efficiency of the production process.

Manufacturers can anticipate fundamental events and take preemptive measures to expect events by utilising AI and machine learning (ML) to analyse real-time data from machine sensors. Smart manufacturing plants utilise exchange rules and machine learning models for determined checking of sensor data, giving bits of information into equipment and handling statuses. This data makes a difference in making qualified choices concerning equipment, shapes, things, operations, clients, and bargains. Smart manufacturing finds application over distinctive businesses, tallying semiconductors, contraptions, remedial contraptions, cars, airplanes, manufacturing equipment, pharmaceuticals, chemicals, metals, mining, and client bundled goods.

Intelligent Manufacturing Systems (IMS) show up in diverse centers of charmed to creators, counting the capacity to expect and address issues, advance quality and reduce downtime, and development by and expansive gear reasonability (OEE).

DOI: 10.1201/9781032630748-6

Computerised eras of era lines energise the amusement of unused making shapes and recognizable affirmation of bottlenecks. Proactive supply chain changes, intelligent stock organisation, and optimisation of coordinations operations, such as bundling and shipping, are empowered. IMS also unveil show day trade openings, compensation streams, and resource monetization lanes, guaranteeing a constant competitive edge. Automation, arranging, and prescient back for thing disillusionments offer assistance in keeping up a vital separate from downtime. Real-time data dealing with and examination at the point of period engage speedy responses to handle anomalies.

Intelligent creating makes a difference for businesses in understanding promote designs and client slants, optimising supply chains by assessing ask, optimising stock, and checking suppliers in real-time. With the coming of the Internet of Things (IoT) and 5G frameworks, businesses can collect and analyse data from a multitude of contraptions and sensors, progress by updating operational, optimization and decision-making capabilities.

However, IMS display new security perils and vulnerabilities. Extended organise and complexity make IMS beneficial targets for cyberattacks. Other than, the tricky data collected and put absent by IMS render them locks in targets for attackers.

Smart mechanical offices epitomise how the Industrial Internet of Things (IIoT) is reshaping standard manufacturing. Manufacturing organisations are getting to be continuously commonplace with adroit generation lines, their capabilities, benefits, and challenges. The move towards mechanical movements like smart generation lines require impressive hypotheses, with thoughts on optimising returns on hypothesis being principal. Respectability must reassess security, vulnerabilities, and other computerised data threats and threats. The conceivable results of assaults run from repercussions to physical wounds, monetary difficulties and decreased efficiency. Frameworks, existing vulnerabilities, planned cyber dangers, deficiencies in current protections, levels of readiness for future security challenges, and the basic need of prioritising security as a foundation in the advancement of future intelligent manufacturing systems. Cyberattacks may nullify the preferences of a savvy production line, counting real-time information observing, supply chain administration, and prescient upkeep. Thus, it is necessary for producers to secure their IMS by installing reasonable security controls for information, systems, gadgets, and applications. Furthermore, cultivating a culture of security mindfulness and preparing users is essential. This chapter explores the security issues and vulnerabilities of IMS in detail. It will also discuss best practices for securing IMS [1–3] (Figure 6.1).

6.2 ARCHITECTURE OF AN INTELLIGENT MANUFACTURING SYSTEM

The integration of AI into shrewdly fabricating essentially happens through intelligent manufacturing systems, speaking to the most unmistakable application. Extending AI past these frameworks may compromise cohesion and viability. These frameworks display particular characteristics inside the "Web additionally AI" system, including independent brilliantly detecting, organising, collaborating, learning,

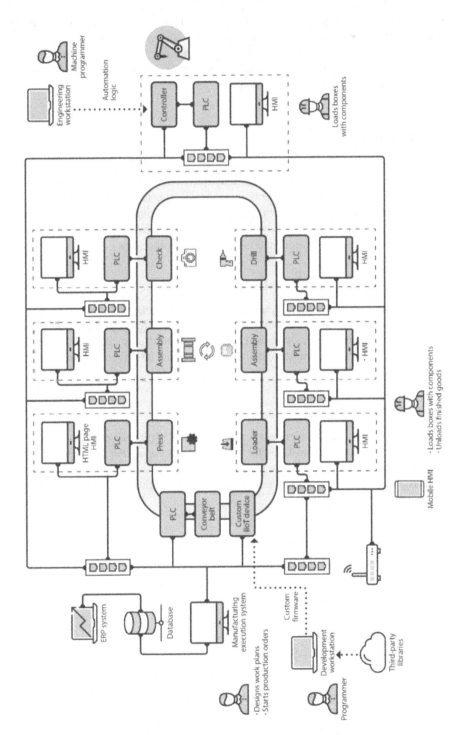

FIGURE 6.1 Intelligent manufacturing systems.

examining, understanding era, decision-making, and execution over human substances, machines, materials, situations, and data frameworks all through their whole life cycles. Regularly, these frameworks contain a few layers, counting an asset and capacity layer, a comprehensive organise layer, a benefit stage, a shrewdly cloud benefit application layer, and a security administration and standard determination system.

6.2.1 RESOURCE/CAPABILITY LEVEL

This layer includes fabricating resources and capabilities, counting physical resources such as machine instruments, mechanical autonomy, and crude materials, as well as intangible resources such as computer program applications and information enbedded in fabricating forms. It moreover includes progressed, advanced, interconnected, and intelligent manufacturing capabilities utilised for different purposes such as show, plan, fabricating, recreation, and integration into processes.

6.2.2 COMPREHENSIVE ORGANISE LAYER

This framework comprises four layers capable of overseeing the network's physical foundation, making and overseeing virtual systems, arranging the organise to meet commerce needs, and checking and announcing arrange execution. It incorporates the physical, virtual, commerce course of action, and intelligent sensing/access layers.

6.2.3 SERVICE STAGE LAYERS

This layer comprises of three primary components: a virtual layer for brilliantly assets and capacities, a central layer for shrewdly back capacities, and a shrewdly client interface layer. It encourages profiling and fine-tuning of beneficial assets, offers basic middleware capacities, and empowers the creation of customised client environments.

6.2.4 APPLICATION LAYER OF CLEVERLY CLOUD SERVICE

This layer empowers people and businesses to utilise cloud administrations in different ways; encouraging detecting, association, collaboration, learning, investigation, expectation, decision-making, control, and assignment execution.

6.2.5 MANAGEMENT OF SECURITY AND STANDARDISED DETERMINATIONS FOR SECURITY

This perspective includes setting up a self-regulating security assurance framework to verify clients, control asset get-to, and protect information inside the intelligent manufacturing framework. It too includes implementing standardised rules along with rules for innovation utilisation, stage get-to, checking, and evaluation.

Intelligent fabricating frameworks speak to an associated and cleverly fabricating benefit framework established in arrange ubiquity and combination. Consistent coordination of people, machines, items, situations, and data, these frameworks empower intelligent manufacturing and on-demand administrations available from any area, giving assets and capabilities in any case of time and put. By leveraging the web (cloud) and intelligent manufacturing assets and capabilities, an organiseorganised brilliantly fabricating framework consistently integrates individuals, machinery, and products.

Also, the innovation in intelligent manufacturing systems comprises different components, counting common innovation, innovation for the intelligent manufacturing stage, common organise innovation, and item life cycle brilliantly fabricating innovation. These components include assorted innovations such as AI, large amounts of information, cloud computing, and rising fabricating advances, which collectively upgrade fabricating prepare effectiveness, quality, security, and development over industries.

The assessment of AI's comprehensive utilisation in intelligent manufacturing includes surveying innovation application, industry suitability, and coming about application results. This assessment incorporates foundation advancement, application ranges, industry improvement, and application viability, centering on upgrading efficiency, capacity, and socio-economic benefits inside the industry [4,5].

The evaluation of AI's comprehensive utilisation in the context of intelligent manufacturing involves an assessment across three primary dimensions: technology application, industry relevance, and resulting application outcomes.

In terms of application technology, it becomes essential to assess the extent and capability of infrastructure development, distinctive applications, general application areas, and business growth. The evaluation of industry development encompasses smart products, which are capable of performing tasks autonomously, and smart connection products that can establish ecological networks. It also includes the evaluation of intelligent industrial software, hardware development that supports intelligent design, manufacturing, management, operation, and safety. Additionally, it involves assessing the development and operation of smart production systems at various levels within the realm of intelligent manufacturing, spanning intelligent manufacturing units, smart workshops, smart factories, and smart industries.

When evaluating application effectiveness, the focus should be directed towards analysing changes in competitiveness and socio-economic benefits. This entails measuring the direct or indirect impact of intelligent manufacturing systems on improving productivity, capacity enhancement, and economic advantages within the industry [6,7] (Table 6.1).

6.3   DATA AND ITS SECURITY

The Internet of Things (IoT) technology has ushered in the Fourth Industrial Revolution (Industry 4.0) by linking factories and businesses to the Internet. "Smart factories" are interconnected with the supply chain via the Internet. Data is transmitted from machines and factories to the cloud, facilitating the exchange of information

TABLE 6.1
Supporting Technologies

Technology Category	Subcategories
General Technology	Intelligent Manufacturing Architecture Technology, Software-Defined Network (SDN) System Architecture Technology, Space-Ground System Architecture Technology, Business Model and Business Modeling Technology, Simulation of Intelligent Manufacturing Services, System Development and Application Technology, Intelligent Manufacturing Safety Technology, Intelligent Manufacturing Evaluation Technology, Intelligent Manufacturing Standardization Technology
Intelligent Manufacturing Platform Technology	Big Data Networking Technology towards Intelligent Manufacturing, Smart Resource/Capacity Sensing Technology and Internet of Things (IoT), Service Environment and Industrial Virtualization Resource Technology/Smart Capacity, Construction Technology/Management/Operation/ Smart Service Environment Assessment, Smart Technology Knowledge/Model Management and Big Data Analytics Technology, Intelligent Human-Interaction Machine Technology/Swarm Intelligence Design Technology, Intelligent Design Technology based on Big Data and Knowledge, Human Hybrid Production Technology, Intelligent Computer Technology, Intelligent Virtual-Physical Combination Testing Technology, Intelligent Production Management Technology for Automatic Decision Making, Intelligent Assurance Technology for Online Remote Support Services
Common Network Technology	Integrated Aggregate Network Technology, Space-Air-Ground Network Technology
Product Life Cycle Intelligent Manufacturing Technology	Intelligent Cloud Innovation Design Technology, Intelligent Cloud Product Design Technology, Intelligent Cloud Manufacturing Equipment Technology, Intelligent Cloud Operation and Management, Intelligent Cloud Simulation and Testing Technology, Intelligent Cloud Service Assurance Technology
Supporting Technology	AI 2.0 Technology, Information and Communication Technology (e.g., Big Data-Based Technology, Cloud Computing Technology, Modeling and Simulation Technology), New Production Technologies (e.g., 3D Printing Technology, Electrochemical Machining Technology), Professional Technology in the Field of Manufacturing Applications (Specialised Technology in Aviation, Aerospace, Shipbuilding, Automobiles, and Other Industries)

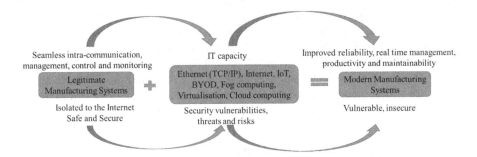

FIGURE 6.2   The transition from legal to modern manufacturing systems.

among all supply chain stakeholders. This enables synchronised and coordinated control over all production aspects, resulting in a simpler, more intelligent, productive, and prosperous manufacturing process life cycle with insights and real-time management.

However, the expanding use of the Internet and mobile devices also exposes modern manufacturing systems to vulnerabilities and threats (Figure 6.2).

6.4   INTELLIGENT MANUFACTURING SYSTEMS DATA

Intelligent Fabricating Systems (IMS) serve as information storehouses, comprising different sorts of significant data for fabricating processes:

- Generation Information: This category incorporates subtle elements concerning fabricating forms, such as hardware status, item quality, and stock levels.
- Client Information: This involves data around clients, counting their buy history, contact subtle elements, and preferences.
- Provider Information: This category includes data with respect to providers, such as contact data, estimating, and conveyance schedules.
- Budgetary Information: This incorporates data relating to the company's monetary execution, such as deals, figures, costs, and profits.
- Mental Property Information: This comprises data concerning the company's items, forms, and technologies.

6.5   SAFEGUARDING MODERN MANUFACTURING
SSTEMS THROUGH CYBERSECURITY

Contemporary industrial and commercial facilities face a plethora of cyber threats, including cybercrime, interference, disruption, data theft, espionage, and denial of process control. These vulnerabilities are exacerbated by the increasing reliance on the Internet and mobile devices in modern manufacturing systems. The automotive industry is particularly susceptible, accounting for nearly 30 percent of all attacks in

2015. Notable incidents, such as the Stuxnet worm attack on Iran's uranium enrichment infrastructure in 2010, and the hacking of a German steel mill furnace in 2014, highlight the severity of the threat. To safeguard against these risks and capitalise on Industry 4.0, businesses must update their security measures. Traditional security tools remain essential but should be augmented with innovative defense methods, such as incorporating security measures into application programs (Security by Design) [8].

6.6   NETWORK AND ITS SECURITY

The integration of new-generation information technology, particularly intelligent manufacturing systems, with the Internet offers unprecedented opportunities for modernising the global manufacturing industry. However, the high connectivity of industrial control systems (ICS) poses significant security challenges, with communication security emerging as a major threat. The absence of standardised protocols for benchmarking design complicates efforts to ensure network security for ICS. Key components of the ICS network include Programmable Logic Controllers (PLCs), Remote Terminal Units (RTUs), Intelligent Electronic Devices (IEDs), and Human Machine Interfaces (HMIs). Different network topologies may be necessary depending on specific ICS application scenarios. Similar work on network technologies is discussed in [9,10].

6.7   DATA SECURITY AND PRIVACY

Specialised manufacturing technologies utilised in industries such as aviation, aerospace, shipbuilding, and automobiles play a crucial role in producing complex and precise parts critical for product performance and safety. The Model-as-a-Service (MaaS) model offers a solution to enhance data privacy protection by allowing data owners to fine-tune global models locally without uploading sensitive data to the cloud. Secure encrypted communication technologies and access control mechanisms further bolster data security and privacy, facilitating digitalisation and intelligence in the manufacturing industry [11].

6.8   IMPACT OF INTELLIGENT MANUFACTURING
SYSTEMS ON SUPPLY CHAIN

Intelligent manufacturing systems are reshaping how companies manage their core functions, including supply chain management. Technologies such as the Internet of Things (IoT), Artificial Intelligence (AI), big data analytics, machine learning, automation, cloud computing, blockchain [12–13], and 3D printing are transforming supply chain operations. Digital transformation can significantly improve efficiency and profitability, leading to cost reductions, increased sales opportunities, and optimised inventory requirements. By leveraging data-driven decision-making, enhancing connectivity and collaboration, and implementing intelligent supply chain practices,

TABLE 6.2
Impact of Intelligent Manufacturing Systems on Supply Chain.

Basis	Key Points in IMS Impact on Supply Chain Efficiency	Remark
Security	Greater transparency and accuracy	Greater transparency and accuracy
Data-Driven Decision	Data-driven decision-making saves costs	Data-driven decision-making saves costs
Connectivity and Collaboration	Enhanced connectivity and cooperation	Enhanced connectivity and cooperation
Inventory Management	Improved warehouse management	Improved warehousemanagement
Adaptive Supply chain	Intelligent supply chain	Intelligent supply chain

IMS enhances supply chain efficiency, enabling businesses to operate more swiftly, flexibly, and accurately [14] (Table 6.2).

6.9 AUTHORIZATION AND AUTHENTICATION IN IMS

Intelligent manufacturing leads to improvements in productivity and product quality, but it also presents notable security concerns. It is concerning to note that almost 80 percent of hacking attacks on manufacturing operations stem from compromised

TABLE 6.3
Opportunities in Intelligent Manufacturing

Opportunity	Description
Increased Efficiency	Intelligent manufacturing systems can significantly help by production efficiency by optimizing processes and reducing wastage.
Cost Reduction	Automation and predictive maintenance can lead to cost savings in terms of labor, energy, and materials.
Quality Enhancement	Real-time monitoring and data analysis can result in better product quality and reduced defects.
Supply Chain Optimization	Integration of supply chain data can lead to better demand forecasting, inventory management, and logistics.
New Business Models	Intelligent manufacturing opens doors for new revenue streams, such as selling data insights or offering remote monitoring services.
Improved Decision Making	Data-driven insights enable better decision-making, helping companies respond quickly to changing market conditions.
Customization and Personalization	Intelligent manufacturing allows for mass customization, catering to individual customer needs.

credentials. Furthermore, it's important to acknowledge that 66 percent of incidents affecting consumer goods and 53 percent of pharmaceutical companies have reported instances of data leaks. Unfortunately, these figures are not unexpected. Compounding the security concerns are the weaknesses in production authentication and access controls; a concerning 44 percent of manufacturers allow unrestricted access to sensitive files for all employees, potentially turning any compromised account into a gateway for widespread data theft.

Faced with escalating threat levels and the growing frequency of ransomware attacks, which can lead to complete shutdowns and substantial financial losses, manufacturing companies must take proactive steps to strengthen their defenses.

It is crucial to minimise attack surfaces and mitigate organizational risk. Considering that logins and authentications represent a critical vulnerability in security, as demonstrated by incidents like the Colonial Pipeline breach initiated by exposed passwords, it becomes essential for security measures to withstand unauthorised attempts to access systems and networks [15].

The following measures can be implemented to enhance the authentication system:

1. Prioritise comprehensive employee training.
2. Implement single sign-on capabilities across disparate networks.
3. Utilise multi-factor authentication for added security layers.
4. Employ secure authentication factors to enhance protection.
5. Enhance user experience to encourage adoption and compliance.

To ensure the security, privacy, and efficiency of an Industrial Internet of Things (IIoT) system, adherence to the outlined design requirements is imperative (Figure 6.3).

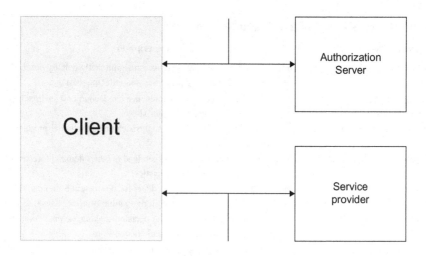

FIGURE 6.3   Authorization and authentication.

6.10   IMPACT OF HUMAN ERROR ON IMS

Every individual is prone to errors. A notable revelation from IBM's "2014 Cybersecurity Intelligence Index" is that 95 percent of security incidents stem from human mistakes. A significant portion of these successful security breaches is orchestrated by external attackers exploiting human vulnerabilities to infiltrate sensitive data.

6.10.1   CYBERSECURITY RISKS

Human mistakes inside Intelligent Manufacturing System (IMS) can posture cybersecurity dangers. Workers might accidentally press on malevolent joins or share sensitive data with unauthorised parties, possibly compromising the security of fabricating forms and data.

6.10.2   OPERATIONAL CHALLENGES

Human mistakes in the information section or handle execution can disturb fabricating operations. Botches in arranging or working apparatus can lead to generation delays, absconds, and quality issues, straightforwardly influencing generation proficiency and item quality.

6.10.3   DATA PRECISION AND QUALITY

IMS intensely depend on precise information for decision-making and prepare optimisation. Human mistakes in information input can result in wrong information examination, affecting the execution of cleverly fabricating frameworks. Destitute information quality can lead to imperfect choices and diminished efficiency.

6.10.4   COMPLIANCE AND ADMINISTRATIVE RISKS

Manufacturing frequently includes compliance with different directions and industry benchmarks. Human blunders in record-keeping or compliance methods can result in administrative infringement, fines, and legal issues. Guaranteeing compliance in IMS requires minimising the chance of human error.

6.10.5   MITIGATION STRATEGIES

While human blunder is unavoidable, IMS can execute relief procedures such as representative preparing, prepare computerization, and real-time observing to diminish its affect. Preparing programs can raise mindfulness of cybersecurity dangers and best hones. Computerisation can minimise manual forms inclined to blunders, and real-time checking can distinguish and address issues instantly. In conclusion, human mistakes can pose critical challenges to IMS, affecting cybersecurity, operations, information quality, compliance, and more. Executing viable relief techniques is significant to minimising the effect of human error and improving the execution and security of intelligent manufacturing systems [16].

6.11 APPLICATIONS OF IMS

Intelligent Manufacturing Systems (IMS) have picked up noticeable quality due to the broad selection of the Web, the expansion of sensors, the development of enormous amounts of information, and great strides in innovation. IMS discover applications in different spaces, leveraging cutting edge data and communication innovation, cleverly science, and progressed generation procedures. A few key applications include:

6.11.1 SMART CITIES

IMS contribute to the improvement of keen cities by optimising forms related to framework, vitality administration, transportation, and open administrations. Savvy city activities depend on IMS for effective asset utilization and moved forward urban living.

6.11.2 SMART HEALTHCARE

In the healthcare segment, IMS play a significant part in improving quiet care through robotization, information analytics, and shrewdly gadgets. This incorporates further understanding the checking, prescient diagnostics, and personalised treatment plans [17].

6.11.3 SMART TRANSPORTATION

IMS are instrumental in the improvement of shrewd transportation frameworks, counting independent vehicles, activity administration, and coordinations optimisation. These frameworks point to progress security, effectiveness, and supportability in transportation.

6.11.4 SMART LOGISTICS

Logistics and supply chain administration advantage from IMS by optimising stock, directing, and conveyance. Real-time information investigation and following improve the proficiency of supply chain operations.

6.11.5 SMART ROBOTICS

IMS drive progressions in mechanical autonomy, making robots more intelligent, more versatile, and competent of errands extending from fabricating to healthcare help. Collaborative robots (cobots) are an outstanding application [18].

6.11.6   SMART COMMUNITIES

IMS back the advancement of keen communities, cultivating maintainability, asset administration, and citizen engagement. These frameworks empower proficient asset allotment and advance the quality of life in communities.

Smart Economy
The integration of IMS into the economy comes about in a savvy economy characterised by effective generation, asset utilization, and data-driven decision-making. IMS contribute to financial development and competitiveness. In general, IMS speak to a transformative drive over different spaces, empowering mechanisation, data-driven experiences, and cleverly decision-making. These frameworks optimise forms, upgrade effectiveness, and drive advancement, contributing to the improvement of savvy and feasible solutions [19].

6.12   CURRENT ADVANCEMENT IN INTELLIGENT MANUFACTURING

Intelligent manufacturing is a quickly advancing field that has earned noteworthy consideration from creative nations. Key improvements and key activities include:

6.12.1   UNITED STATES INITIATIVES

In the United States, innovative strategies and policies have been introduced, such as the "Advanced Manufacturing Collaboration Plan" (2011) and the "Internet Industry" (2012). In 2012, General Electric (GE) introduced the concept of the "Industrial Internet," aiming to connect intelligent equipment, people, and data for more informed decision-making. The Industrial Internet comprises smart devices, smart systems, and smart decision-making, enabling data analysis and visualisation for intelligent decisions.

6.12.2   GERMANY'S INDUSTRY 4.0 PLAN

Germany initiated the Industry 4.0 plan in 2013, emphasising "smart factories" and "Intelligent manufacturing" as primary research areas. The plan introduced the strategic concept of "one core, two subjects, three-dimensional, and eight-plane integration." Germany has also developed digital cloud service platforms like Siemens' Scialytic.

6.12.3   CHINA'S TRANSITION AND STRATEGIC PLANS

China is undergoing a historic transition from lower-value manufacturing to higher-value stages of the value chain. Key strategic plans include "Made in China 2025," "State Council Guidance on Promoting Internet+ Action," and more. "Made in

China 2025" outlines a 30-year plan focused on innovation, quality, sustainability, and talent development. China aims to become an innovation-driven powerhouse.

6.12.4 CHINESE ADVANCEMENTS IN TECHNOLOGY

China has made significant advancements in network infrastructure, high-performance computing, network communication devices, and intelligent software. This has led to developments in mobile Internet, big data, cloud computing, and smart devices. For example, the global market for wearable smart devices has seen substantial growth, with China making significant contributions to this market. Overall, intelligent manufacturing is a strategic focus for many countries, with China actively transitioning into a leader in this field, driven by advancements in technology and strategic plans to boost innovation and economic benefits.

6.13 OPPORTUNITIES IN INTELLIGENT MANUFACTURING

Intelligent manufacturing systems present significant market opportunities and are being widely adopted in production systems worldwide. According to a report by McKinsey and Company presented by Manyika, the accommodation and food services sector have a 73 percent potential for automation, while the manufacturing, transportation, and warehousing sector have a 60 percent automation potential. Given the high automation potential in these sectors, extensive research is currently focused on deploying cutting-edge technologies to enhance productivity and efficiency. The 2017 WBR (Worldwide Business Research USA, LLC) survey report has revealed that many survey participants have plans to upgrade their production systems. Among them, 38 percent intend to install new technology, 45 percent plan to enhance existing machinery with intelligent manufacturing systems, and 17 percent are considering a combination of both approaches. Current manufacturing systems often lack critical components and functionalities found in intelligent manufacturing technologies. Key features of intelligent manufacturing systems, such as self-configuration, self-optimization, early perception, decision-making, and predictive maintenance, are often missing in even the most advanced manufacturing systems, including reconfigurable production systems. The architecture of an industrial automation system is determined by the complexity of the system, the interactions among machines, the roles of subsystems, and their working mechanisms and dependencies. Automation technology has enabled the sharing of information and data across interconnected systems, rapidly increasing process management and interoperability between machines. This facilitates flexible resource and process management and effective exception handling. The adoption of automated and intelligent systems has ushered in the era of intelligent manufacturing, which leverages end-user data and cloud computing to accelerate the production of customised products. Strong robotisation frameworks have become a need to meet the requests of today's advertise, optimising generation costs, upgrading plan realness, and diminishing item conveyance times. Intelligent fabricating is not an opportunity but a necessity to flourish in the ever-evolving scene of advanced fabricating [20].

6.14 CONCLUSION

In conclusion, the vision of Intelligent Manufacturing System (IMS) paints a picture of a future where machines work consistently, fueled by Artificial Intelligence, the Internet of Things (IoT), and enormous information analytics. IMS is reshaping the fabricating scene by optimising effectiveness, efficiency, quality, and supportability. It brings various benefits like real-time information checking, prescient support, and progressed supply chain management.

However, this innovative headway, too, brings along noteworthy security challenges. The increased network and complexity of IMS make them prime targets for cyberattacks. The sensitive information amassed by IMS, extending from generation subtle elements to client profiles and mental property, makes them luring targets for cybercriminals. Thus, shielding IMS is foremost to opening their full potential.

To reinforce security, organizations must center on worker preparation, execute vigorous confirmation conventions, and consider embracing multi-factor confirmation. Additionally, the effect of human blunder in cybersecurity cannot be neglected, requiring the utilisation of innovation to relieve related risks.

IMS has a significant effect on the supply chain, cultivating more prominent straightforwardness, data-driven decision-making, upgraded network, and progressed stockroom administration. Whereas transitioning to advanced, computerised supply chains pose critical challenges, the benefits in terms of effectiveness and taken a toll lessening are substantial.

Furthermore, the possibilities displayed by IMS are tremendous, especially in segments with multi-mechanization potential such as fabricating, transportation, and warehousing. These innovations empower self-configuration, self-optimization, prescient upkeep, and more, eventually driving efficiency and efficiency.

In summary, IMS stands as a transformative drive in fabricating and promoting a pathway to increased proficiency and competitiveness. However, the appropriation of IMS must go hand-in hand-with strong cybersecurity measures to counter developing dangers. As organisations explore this mechanical move, they must prioritise security, contribute to its preparation and use innovation to moderate human error, all while seizing the colossal potential that intelligent manufacturing presents.

REFERENCES

1. Alex Khang, et al. "Enabling the future of manufacturing: integration of robotics and iot to smart factory infrastructure in industry 4.0." In *Handbook of research on AI-based technologies and applications in the era of the metaverse.* IGI Global, 2023. 25–50.
2. Y. C. Ahmad Barari, Marcos de Sales Guerra Tsuzuki and M. Macchi, *"Editorial: intelligent manufacturing systems towards industry 4.0 era,"* Springer, 2021.
3. Adib Habbal, Mohamed Khalif Ali, and Mustafa Ali Abuzaraida. "Artificial intelligence trust, risk and security management (AI TRiSM): Frameworks, applications, challenges and future research directions." *Expert Systems with Applications* 240 (2024): 122442.

4. Dennise Mathew, N. C. Brintha, and J. T. Winowlin Jappes. *"Artificial intelligence powered automation for industry 4.0."* New Horizons for Industry 4.0 in Modern Business. Springer International Publishing, 2023. 1–28.

5. Q. P. Yang Li, and Shihao Wu, "Network security in the industrial control system: A survey," *arXiv:2308.03478*, 2023.

6. Chao Zhang, et al. "Towards new-generation human-centric smart manufacturing in Industry 5.0: A systematic review." *Advanced Engineering Informatics* 57 (2023): 102121.

7. R. B. Sudip Phuyal, and Diwakar Bista, "Challenges, opportunities and future directions of smart manufacturing: A state of art review," *Sustainable Futures* 2 (2020), https://doi.org/10.1016/j.sftr.2020.100023

8. Li Yang, et al. "Adoption of information and digital technologies for sustainable smart manufacturing systems for industry 4.0 in small, medium, and micro enterprises (SMMEs)." *Technological Forecasting and Social Change* 188 (2023): 122308.

9. M. A. Ali, B. Balamurugan, R. K. Dhanaraj, and V. Sharma, "IoT and blockchain based smart agriculture monitoring and intelligence security system," *2022 3rd International Conference on Computation, Automation and Knowledge Management (ICCAKM)*, Dubai, United Arab Emirates, 2022, pp. 1–7, https://doi.org/10.1109/ICCAKM54721.2022.9990243

10. A. Raj, V. Sharma and A. K. Shanu, "Comparative Analysis Of Security And Privacy Technique For Federated Learning In IOT Based Devices," 2022 3rd International Conference on Computation, Automation and Knowledge Management (ICCAKM), Dubai, United Arab Emirates, 2022, pp. 1–5, https://doi.org/10.1109/ICCAKM54721.2022.9990152

11. M. A. Ali, B. Balamurugan, V. Sharma. "IoT and blockchain based intelligence security system for human detection using an improved ACO and heap algorithm." *2022 2nd International Conference on Advance Computing and Innovative Technologies in Engineering (ICACITE)*. IEEE, 2022.

12. A. Raj, V. Sharma, S. Rani, A. K. Shanu, A. Alkhayyat and R. D. Singh, "Modern farming using IoT-enabled sensors for the improvement of crop selection," *2023 4th International Conference on Intelligent Engineering and Management (ICIEM)*, London, United Kingdom, 2023, pp. 1–7, https://doi.org/10.1109/ICIEM59379.2023.10167225

13. T. Hai, et al. A novel & innovative blockchain-empowered federated learning approach for secure data sharing in smart city applications. In: Iwendi, C., Boulouard, Z., Kryvinska, N. (eds) *Proceedings of ICACTCE'23 — The International Conference on Advances in Communication Technology and Computer Engineering. ICACTCE 2023. Lecture Notes in Networks and Systems*, vol 735. Springer, Cham, 2023. https://doi.org/10.1007/978-3-031-37164-6_9

14. V. Juyal, R. Saggar, and N. Pandey, "Clique-based socially-aware intelligent message forwarding scheme for delay tolerant network," *International Journal of Communication Networks and Distributed Systems* 21 (2018): 547–559.

15. A. Singh, R. K. Dhanaraj, M. A. Ali, B. Balusamy and V. Sharma, "Blockchain technology in biometric database system," *2022 3rd International Conference on Computation, Automation and Knowledge Management (ICCAKM)*, Dubai, United Arab Emirates, 2022, pp. 1–6, https://doi.org/10.1109/ICCAKM54721.2022.9990133

16. N. Manikandan, D. Ruby, S. Murali, and V. Sharma, "Performance analysis of DGA-driven botnets using artificial neural networks," *2022 10th International Conference on Reliability, Infocom Technologies and Optimization (Trends and Future Directions) (ICRITO)*, 2022, pp. 1–6, https://doi.org/10.1109/ICRITO56286.2022.9965044

17. V. Juyal, R. Saggar, and N. Pandey, "On exploiting dynamic trusted routing scheme in delay tolerant networks," *Wireless Personal Communications* 112 (2020): 1705–1718.
18. K. Durgalakshmi, P. Anbarasu, V. Karpagam, A. Venkatesh, B. Kannapiran, and V. Sharma, "Utilization of reduced switch components with different topologies in multi-level inverter for renewable energy applications-a detailed review," *2022 5th International Conference on Contemporary Computing and Informatics (IC3I)*, Uttar Pradesh, India, 2022, pp. 913–920, https://doi.org/10.1109/IC3I56241.2022.10073430
19. N. Pushpalatha, S. Jabeera, N. Hemalatha, V. Sharma, B. Balusamy, and R. Yuvaraj, "A succinct summary of the solar MPPT utilizing a diverse optimizing compiler," *2022 5th International Conference on Contemporary Computing and Informatics (IC3I)*, Uttar Pradesh, India, 2022, pp. 1177–1181, https://doi.org/10.1109/IC3I56241.2022 .10072844
20. V. Sharma, N. Mishra, V. Kukreja, A. Alkhayyat and A. A. Elngar, "Framework for evaluating ethics in AI," *2023 International Conference on Innovative Data Communication Technologies and Application (ICIDCA)*, Uttarakhand, India, 2023, pp. 307–312, https://doi.org/10.1109/ICIDCA56705.2023.10099747

7 Multi-Factor Authentication Methods in Intelligent Systems

Prachi Dahiya and Umang Kant

7.1 SECURITY ISSUES IN INTELLIGENT SYSTEMS

Industry 4.0, a conglomeration of diverse trends and technologies that has fundamentally altered the way commodities are produced in an industrial setting, heavily relies on computer science. Due to the advanced systems, security is further strengthened for a multitude of cyberthreats. The kind of industrial transformation that manufacturers are operating in is giving rise to a variety of new industries. Since these intelligent systems are a natural venue for meetings, computer science has made considerable gains in developing strategies and leveraging new technology for these platforms [1]. In the event that fresh information becomes available and drastically modifies the regulations, it is essential to closely monitor the current developments.

The digitalization of Industry 4.0 has made security an increasingly pressing issue for the industrial sector. The rise in security concerns associated with Industry 4.0 can be attributed to several factors, including the limitations of computing power, the sharp increase in the amount of data exchanged over networks for sophisticated frameworks, problems with availability across many business sectors, and the demonstration of remarkable failure power in wide-region organizations. With the proliferation of computer science tools, smart systems are beginning to resemble companies. The novel idea of human-machine connection is difficult for factory workers to understand, which causes issues with augmented reality systems or subpar system interaction interfaces.

The views of the industrial environment have been altered as a result of these new digital trends, and a lot of research is being done in a number of application areas that are not yet in use. A smaller number of technologies provide higher reliability and lower application area costs, but more research is being done these days to improve the quality of industrial applications. Many industrial dealers are still unaware of recent developments in computer science technologies that are used in intelligent systems, such as machine learning and the Internet of Things.

Attackers are focusing on Industry 4.0 as security flaws become more prevalent with each new technology. Numerous hostile acts in a range of Industry 4.0 industrial applications can give rise to security concerns. Malicious actors use a variety of tactics, including industrial espionage, property theft, intellectual property theft,

DOI: 10.1201/9781032630748-7

disclosing private information, and sabotage. Industry 4.0 is now the second most attacked industry, primarily due to security flaws in intelligent systems.

The factories that are part of Industry 4.0 manufacturing systems are known as intelligent factories. Denial of service attacks, device exploitation, information sharing vulnerabilities, hacking of networked devices that exchange data and use it to control devices, and various other types of attacks that can occur in systems associated with networks are the main issues a company deals with. It could be difficult for manufacturers to recognize and defend against cyberattacks of any kind. The proliferation of IoT networks in production systems has led to a regular occurrence of significant cyberattacks [2]. Figure 7.1 illustrates the problems identified at different points during the industrial revolution and the approaches that new-age systems may need to take in order to deal with any attacks.

Every networked part of the manufacturing system puts the environment of the network at risk and in danger. Industry Control Systems (ICS) have a number of vulnerabilities that could make them more vulnerable to hackers. In the industrial sector, smart machines and gadgets have replaced older, split equipment and gadgets because they enhance the perimeter for prospective attackers. Furthermore, these gadgets are difficult to upgrade because the programmers or other specialists in charge of the updates must make several alterations that the production staff does not comprehend. The extremely limited number of regulatory requirements in industrial activities is difficult for employees to understand. Communication can occasionally be challenging since many environments operate as separate, autonomous entities [3].

Basic defensive tactics are among the risk-based security measures that the industrial sector must implement. Information technology (IT) and operational technology (OT) assets need to be safeguarded using a certain set of security protocols. Secure data and folders, out-of-date systems, and other problems can all be successfully patched with integrated defensive techniques across several platforms. The goal of security measures should be to identify the threat before it does significant harm to network-connected devices. It is imperative for vendors and manufacturers to ensure that security software patches are updated regularly to ensure that devices remain free of faults. This will stop any breach before it happens and allow for the implementation of the necessary safety measures to protect the devices and data.

7.1.1 CYBER PHYSICAL SYSTEMS (CPSs)

Cyber physical systems consist of various pieces of machinery, gear, digital networks, and hardware that are combined into functional systems. All of these gadgets are cyber physical systems with network environment operation, communication, and collaboration capabilities. They are all related to machine learning, computer systems, and computational intelligence. The creation of the Internet of Things by the computer science community has considerably helped the world and Industry 4.0. Significant progress in Industry 4.0 has been made possible by a multitude of technologies, including deep learning, machine learning, artificial intelligence, and big data approaches. These developments have led to the development of intelligent

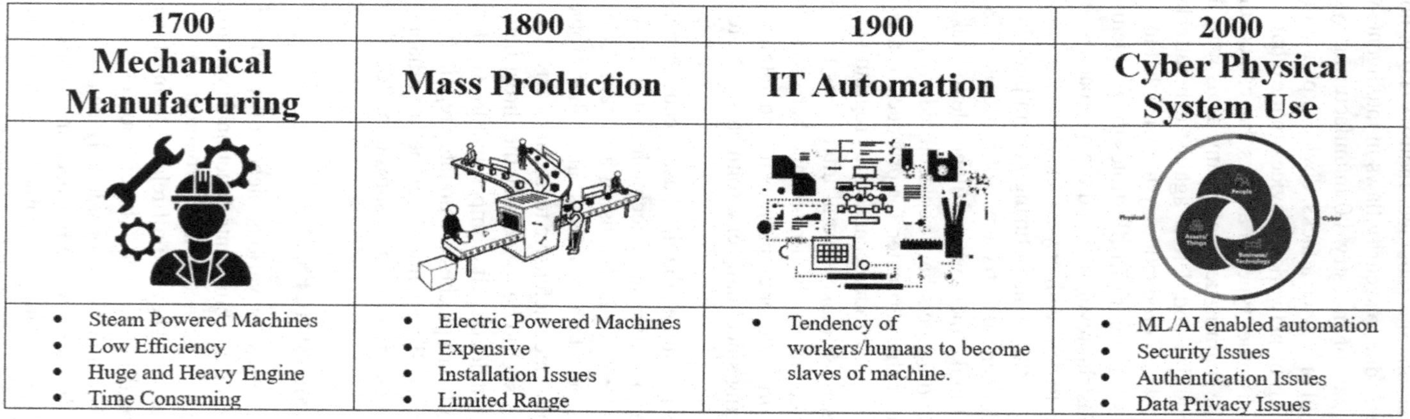

1700	1800	1900	2000
Mechanical Manufacturing	**Mass Production**	**IT Automation**	**Cyber Physical System Use**
• Steam Powered Machines • Low Efficiency • Huge and Heavy Engine • Time Consuming	• Electric Powered Machines • Expensive • Installation Issues • Limited Range	• Tendency of workers/humans to become slaves of machine.	• ML/AI enabled automation • Security Issues • Authentication Issues • Data Privacy Issues

FIGURE 7.1   Industrial revolution concerns.

systems and smart factories in the sector. Unknown security risks are raised by the integration of informational technologies, including information technology, operational technology, and intellectual property [4].

Systems are susceptible to hostile behavior from both IT and OT, and organized groups of attackers frequently target manufacturing systems with the intention of carrying out a range of damaging deeds. Such hostile activities could be caused by a variety of factors, including industrial espionage, data breaches, production management errors, and system failures. Devices and data are vulnerable in production systems with a high number of digital and physical components. Industrial systems need upgrades to address security flaws and trained security administrators are the ones who can provide these updates. Updates for important functionality in outdated software can occasionally fail to install.

Certain OT systems might not have the security updates for a variety of reasons. To ensure that the production system performs at its peak, it is imperative that these systems be operated with as little interference from external resources as possible. The different interfaces also require updates. Industrial control systems (ICS) and human-machine interfaces (HMI) are usually used at remote locations. These ICS are vulnerable to various cybersecurity threats in the IT and OT domains of industrial systems since they use a variety of networks.

ICS components are vulnerable to a range of security threats since they are widely used in manufacturing systems and are available to the public. Attackers have developed a routine of gathering publicly accessible documents and using them for any purpose they choose over time. This includes malicious activities involving data, hardware, privacy, and products, formulas, machine blueprints, and other functional technologies. Computer-aided designs, or CAD, contain information about the equipment being utilized; if this information is made public, malicious individuals will use it for their own damaging objectives.

Many intelligent systems utilize the computational power of information systems to enhance machinery, business models, and the convergence power of current and emerging technology. Popular fixes for different technical issues, such as "choose your own device" (CYOD), "bring your own device," and "bring your own demon," also introduce new security risks. New and diversified risks will emerge as a result of the ongoing trend of developing new tools and technology. Environments will alter as a result, new regions will expand, and new security solutions will need to be created to deal with these problems. Along with the development of new technologies, there is also an ongoing identification and creation of new technological security implications and solutions. Improved networks, increased production and efficiency, beneficial communications, robust frameworks, and chances for research that will result in better applications are all possible with intelligent machines and intelligent networks. All these acts raise serious concerns about the security of the framework, seriously jeopardize the infrastructure, and have a major and expensive effect on the property and equipment utilized in Industry 4.0. Extended production lines give rise to software vulnerabilities, which can potentially escalate physical attacks and impede system output under certain conditions.

In the industrial sector, the fourth generational revolution began in the 2000s and is expected to expand significantly faster than anyone could have predicted. Some of the technologies that have replaced the antiquated industrial processes are automation, data analytics, robots, and neural networks; these new intelligent systems are currently taking over everywhere [5]. Occasionally, however, the revolution has also been able to go more easily to the next phase thanks to these increasingly perceptive, wise, and efficient technologies. The Industrial 4.0 era's automation and digitization are to blame for the apparent fleeting change.

7.1.2 INDUSTRIAL INTERNET OF THINGS (IIoT)

The term "Internet of Things" (IoT) refers to the networks that numerous industries and enterprises use for data sharing and communication. It has been shown that the development and efficacy of Industry 4.0 depend on these networks. A specific IoT network houses smart sensors, actuators, various devices such as RFID tags, factories and their locations, and their data concepts, and are implemented with the aim of improving, among other things, utilities, management, dependability, productivity, and efficiency. They use information and data for a variety of reasons, analyzing and exchanging it.

These sensors and actuators help to detect inefficiencies or fluctuations in the output or efficiency of the production system, which saves time and energy and helps to promptly and economically detect system problems. These facilitate enhanced traceability of the supply chain, environmentally friendly methods of producing energy, sustainable innovations for Industry 4.0, quality assurance, and general advancements in the production and quality of domains. The deployment of IoT and the provision of suitable and dependable resources are necessary to meet predictive maintenance goals, such as energy management resources and reliable asset tracking. Figure 7.2 provides a detailed overview of the features and elements of the IoT network.

The IoT consists of the following components:

1. A variety of factory devices that function in an environment where data is collected, managed, shared, watched over, analyzed, and transferred amongst them; production data and reports are then shared with various establishment or groups.
2. Confidential information sharing and correspondence inside a comparable organization plan, as well as public information sharing and correspondence among other IoT organizations.
3. The sector makes critical decisions and boosts business efficiency with the use of homemade analytical tools.
4. IIoT data warehouses contain state-of-the-art manufacturing and commercial storage capacities.
5. The IIoT network cannot operate effectively without both machines and people. Both of these are necessary for it to function.
6. Edge devices also make a major contribution to the enhancement of business processes through the provision of business solutions, effective decision-making, optimal machinery operations, proactive maintenance of the devices and data, and so forth.

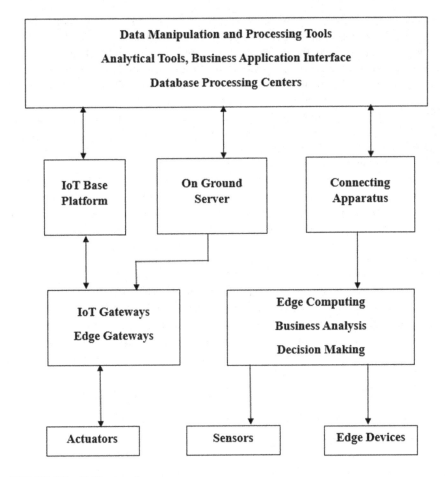

FIGURE 7.2 IIoT network.

7.1.3 ISSUES WITH CYBERSECURITY IN ASTUTE FRAMEWORKS

The main issue in this digital age of Industry 4.0 is illegal access to networked smart devices that act contrary to approved decision-making processes. Due of their easy access to the data on these devices, hackers are able to seriously disrupt production. Typical attacks that could get more intense in this situation include:

1) Malware Interrupted

In recent times, software programmers have discovered methods around firewalls, so the firewalls that are supposed to purge malware from devices are insufficient. In essence, downloading an adverse programming system is insufficient to ensure the device and other linked devices are safe. Interlopers misuse a few vulnerabilities and disparities that exist from where they can damage the whole framework. Examples of the diverse sorts of spyware that operate in certain groups are worms, ransomware, spyware, trojan horses, and so on. These days, more sophisticated versions of spyware are also in use. This might damage the production structure, the functioning of the system as a whole, the mechanical assembly, the store network, and the plant structure, causing chaos and destruction [6]. Progressed levels of the firewall systems are to be employed, and Normal Web Record Structure Sharing Judgment and Noticing is one of the most fundamental ones. The structural climate is validated by this system through unauthorized correspondence, updates, and alterations, among other things. There are a few occasions when this extra security will be necessary, and all the relevant specialists are advised to report any unauthorized access to the system. Additionally, basic robotizations are completed here.

These days, a variety of firmwares are utilized to coordinate and keep developers at bay in the IIoT space. With the increasing number of vulnerabilities, organizations are using increasingly sophisticated and powerful firewalls to ensure that security measures are met even in the event that they become apparent. These security escape statements have the potential to cause the entire symphony to malfunction. Because the great majority of machines have lower levels of affirmation, hackers will find it easier to attack them and steal data from them. These attacks are making the organizations handicapped such that the organizations are not able to do any work securely.

In order to migrate the drivers and provide protection for the most recent virus to appear in the orchestra, many developers upgrade their firewall and program. A client-driven access structure should be set up so that no unauthorized device can access the system in any way, including by destroying the access, restricting the access, entangling the messages in the event that they are sent by a distant gatecrasher, impeding the actions carried out by any unauthorized device, and so on.

2) Worries of BYOD

Bring your claim gadget has gotten to be the conversation of the town but it comes with its claim of security issues, and in this culture, representatives are, in reality, empowered to bring their claim gadgets such as keen phones, portable workstations, desktop screens so that the workers can be more productive and give better outputs. When the workers bring their own portable workstation, they get to be more comfortable and their execution will certainly make strides. The BYOD culture is the major reason of sparing a part of capital sum utilized in Industry 4.0, and this too diminishes the capital expenditure in the company to acquire distinctive gadgets for the workers. If the organize is not scrutinized at that point, the IIoT organize would end up helpless and this would lead to a few other dangers and will disturb the whole workflow. There is a require that each kind of firewall that is required in the framework ought to, as of now, be introduced such that security from each danger can be there.

3) Inaccessibility of Ongoing Encryption

Genuine time encryption is vital at the equipment level of a cleverly framework and in such cases if the encryption is not there at that point, there is defenselessness to the whole framework. All the data is scrambled that is to be shared to keep the organize secure and at that point the data is traded at the granular level such that no programmer is able to assault or spill the data to other noxious frameworks. This is the need of the hour to scramble the data which is traded inside the arrange. If the off-base or noxious information comes up at that point, this leads to genuine harm to the generation levels or can lead to a total shutdown. Savvy encryption arrangements are required along with the optimized operations which can scramble the real-time information.

7.2   MULTI-FACTOR AUTHENTICATION (MFA)

In the present computerized scene, shielding sensitive data is foremost. To invigorate protections and ensure against unauthorized get-to, it's significant to emphasize the utilization of numerous verification variables [7]. Multi-Factor Authentication (MFA) is a fantastic security method that essentially lowers the risk of breaches.

Multi-Calculate Confirmation likewise is a two-step verification (2SV) or two-factor authentication (2FA) system; this security setup requires users to provide at least two separate confirmation pieces in order to access a system, file, or application. The purpose of MFA is to enhance security by adding a second degree of verification beyond a login and password, which are vulnerable to phishing and brute force attacks, among other types of attacks

The three standard categories of verification components utilized in MFA:

Something You Identify: Most of the time, this is the pair of a conventional username and secret password. The first component is something that the user can provide and is aware of.

Something You Own: This element relates to a real object that the customer owns, such as a security key, smart card, equipment token, or mobile phone. The customer should enter the secret word after providing this actual record of themselves in some form.

Something You Possess or Are (Biometric Information): Such as voice, iris, fingerprint, or facial recognition is the subject of this component. Biometrics are an individual's remarkable social or physical attributes that may be used to get approval.

In order to finalize the validation procedure, the customer must normally supply a minimum of two of these factors. For instance, a user may submit their password (something they know) when entering into an online account with MFA enabled, and they will subsequently receive an SMS or dedicated app on their smartphone containing a one-time code (something they have). To get in, they would have to enter the code.

Because an aggressor would ultimately need to have the subsequent element or the biometric traits of the authentic customer in order to arrive at an account, even if they were to manage to acquire or guess the password, multi-factor authentication greatly strengthens account security. As a result, it is far more difficult for unauthorized people to compromise systems and accounts [8]. Various MFAs can be seen in Figure 7.3.

7.2.1   VARIETY OF MFA

The part discusses several MFAs.

FIGURE 7.3 Variety of MFAs.

- Two-Variable Validation (2FA).
- Two separate kinds of authentication from different categories of elements are required, or two forms of two-factor authentication (2FA), in order to verify a customer's identity and prevent unwanted entry into a database, application, or online profile. Typically, these components belong to one of three classes, as was discussed in the preceding section: (i) Something You Know, (ii) Something You Have, or (iii) Something You Are. When two of these components are used in tandem, known as two-factor authentication (2FA), the authentication process is strengthened as opposed to when a password is used alone, which is vulnerable to hacking via phishing, brute force attacks, or credential leaks.

The four main steps in the 2FA process are as follows, as seen in Figure 7.4:

- a. User initiates login: A user enters their username and password as the first factor (something they know) while attempting to access a system or account.
- b. Second-factor request: The user is prompted for a second authentication factor by the system after entering their password. This is typically

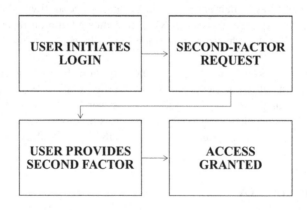

FIGURE 7.4 2FA process.

communicated to a means or device connected to the user, like a hardware token, email, text message (SMS), or smartphone app.

c. Second element provided by the client: The second variable, which may be a one-time code, an output of a single finger imprint, or some other type of ownership- or personality-based validation, is provided by the client.

d. Access granted: The user receives access if both factors are correctly validated. If one of the factors is missing or inaccurate, access is refused.

By making it substantially more difficult for an attacker to get unauthorized access, 2FA dramatically improves security [6]. This is achieved by making the attacker reconsider the client's secret phrase and the next component. 2FA is a feature that many online businesses and apps provide as a way to safeguard user accounts and private data against hacking and illegal access.

7.2.2 THREE-ELEMENT VALIDATION (3FA)

Three-Element Validation (3FA) is a higher level of security that adds another layer of verification to the concept of Two-Component Approval (2FA). Similar to 2FA, 3FA requires users to provide three free factors to validate their identity at any point before granting access to an online record, application, or structure. These elements, which have been covered in previous sections, often fall into the same categories as 2FA: (i) Something You Know; (ii) Something You Have; and (iii) Something You Are. Figure 7.5 shows the methods that are employed in 3FA.

Updating security is undoubtedly the main goal of 3FA in comparison to 2FA. Requiring three distinct elements makes it more difficult for unauthorized users to access a record or structure since they have to consider the three confirmation pieces. Using 3FA might make sense for very sensitive buildings or critical establishments where an extra security measure makes sense. Nevertheless, it's critical to take into account the customer's experience as well as any possible complexities that may occur with more elements. Adjusting security with ease of use is vital to guarantee that 3FA doesn't get to be excessively burdensome for genuine clients. Subsequently,

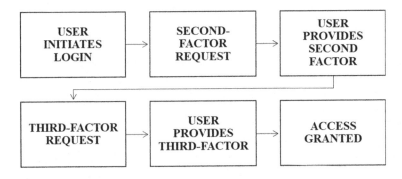

FIGURE 7.5   3FA process.

in hone, 3FA is less common than 2FA since it can be more lumbering for clients and may not continuously be fundamental for all sorts of accounts or frameworks [9]. The choice to implement 3FA must ~~to~~ be based on a careful risk assessment and the particular security needs of the organization or application. Figure 7.5 represents the 3FA handle that was remembered.

- Verification via Biometrics

  A security system known as biometric affirmation uses a person's distinct physical or social characteristics to validate their identity. Unlike traditional methods of verification such as PINs or passwords, which need knowledge, biometric authentication relies on "something you are." Since biometric data is difficult to replicate or obtain, it provides strong evidence for confirmation.

Biometric authentication has four advantages: (i) Security: Biometric data poses significant security issues because it is difficult to produce and obtain, (ii) Convenience: Customers do not need to send physical tokens or repeat passwords, (iii) Speed: Biometric authentication can happen quickly, especially when facial recognition or special reservation is used while waiting for the new version, ~~technologies~~ (iv) Don''t stop: biometric identification can reveal a person's relationship and activities, thus taking on a new role. But there are problems with biometric authentication [10]. When processing biometric information, there are considerations such as security, the possibility of false or inaccurate information, and the need for special equipment or sensors. Similarly, biometric data needs to be carefully monitored to ensure people's trust and compliance with important laws and guidelines.

- Time-Sensitive One-Time Password (TOTP)

Time-Sensitive One-Time Password (TOTP) provides two-factor authentication (2FA) functionality that generates temporary passwords to increase the security of online information and machines. TOTP is often used to provide greater security than traditional passwords.

There is a need to examine the functions of TOTP in detail, which will become clear later.

- Age of the Secret Key

When a client configures TOTP for an online document or application, the user's device and the server generate a secret key, which is also referred to as a shared secret and is maintained securely. A cryptographic value specific to the user and the service is this secret key.

Time-Based Code Generation: The mystery key and the running time are used by the client's device as well as the server to feed a cryptographic method (usually HMAC-based) to generate a one-time secret phrase. This approach generates a six- or eight-digit mathematical code.

Code Legitimacy Window: Depending on how the TOTP is implemented, the generated code is simply available for a short period of time, between 30 and 60 seconds. After this window of opportunity, the code expires.

Client Validation: The user enters the most recent TOTP code that their device has generated in order to log in or finish a transaction. Using the user's secret key, the server independently determines the expected code and compares it to the code that the user sent. Authentication is successful if they coincide and are within the allowed time frame.

Beyond passwords, TOTP adds an added degree of security. Without the TOTP code, a malevolent actor cannot have access to the data, even if they manage to obtain the client's confidential phrase. As a result, security is improved. It is not too difficult to use and set up. In order to generate codes, users usually need an app (such as Google Authenticator or Authy) and a smartphone or other TOTP-compatible device. Because the clocks on the device and the server are synchronized, it functions even when the user's device is not connected. Because it is built on open standards and can be applied to a wide range of services and apps that enable 2FA, it is often performed on a large scale for a great number of clients and services without requiring a substantial amount of infrastructure complexity.

- Sensible Cards and Tokens

Sensible Cards and Tokens serve as tangible objects utilized as a validation factor to bolster safety measures. They play a crucial role in MFA setups by offering an added layer of security beyond conventional passwords. The primary factors to consider when it comes to smart cards and tokens are:

   (i) Cost and deployment: Initial investments in hardware and infrastructure are necessary for smart cards and tokens.
  (ii) When added to the authentication loop, both compromised cards and tokens will affect customers. Proper planning and support are essential to ensure success.
 (iii) Board of Directors and exceptions: In the event of disaster, theft, or employee turnover, the organization must have procedures for rewards, reimbursement, and termination in place. Say no to notes and tokens.
 (iv) Interoperability: Effective use of smart cards and tokens in the MFA environment requires compatibility with existing standards and practices.

7.2.3 PLACE-BASED VERIFICATION

MFA technology, called location authentication, uses the user's current location to verify their identity. This method adds an additional degree of security by using the user's device location as one of the authentication criteria. Here's how location-based authentication functions:

a. Gathering User Location Information: In order to employ location-based authentication, the user's device—a laptop or smartphone, for example,continuously gathers location information using a variety of technologies, such as GPS, Wi-Fi, cellular towers, and IP address geolocation.

b. Request for Validation: The location-based verification system may ask to access the client's device location information when the user tries to access an application or system that needs authentication.

c. Verification: The framework considers the permitted or projected area after analyzing the area information it has received. The verification is considered as successful if the user's device is located inside a predetermined boundary or within an authorized geographic area.

d. Conditional Access: Based on the user's location, access may be allowed or prohibited in accordance with the organization's security regulations. For instance, access might be prohibited if the user's device is not within a designated area.

Utilizing area-based verification has numerous benefits such as:

(i) Improved Security: Location-based verification includes an additional layer of security by guaranteeing that get-to is allowed as it were if the client's gadget is genuinely in an anticipated or approved area. This can obstruct further assailants who might have stolen qualifications, however are not the valid.

(ii) Associations have the ability to implement geofencing, a technology that establishes boundaries on territory that may be accessed. Depending on how close the client is to certain areas, geofencing can be utilized to provide or refuse access.

(iii) Crucial for Hard-to-Reach Work: Since it can help guarantee that customers are accessing company resources from approved regions or locale, area-based affirmation can be particularly helpful in remote work scenarios.

(iv) Consistency: Some organizations and management frameworks may require associations to complete area-based affirmation in order to agree to security and data security standards.

When employing region-based affirmation, there are a number of factors that need to be taken into consideration. There are a few important elements that are taken into account given as follows:

(i) Issues with Insurance: The collection and use of region-specific data presents security challenges. To address customer concerns, associations should have well-defined security strategies and permission forms in place.

(ii) False Positives and Negatives: Area data may be erroneous or prone to errors, which might result in false positives (hindering traditional clients) or false negatives (permitting unauthorized access).

(iii) Parodying and VPNs: To conclude, the aggressors may attempt to appear to be inside the authorized reach by spoofing their location or using virtual private networks (VPNs).

(iv) Client Experience: If a client often travels or works remotely, they may find that an area-based inspection is intrusive or poorly organized.

(v) Backup Instruments: In the event that region data is faulty or missing, associations need to have optional affirmation procedures in place.

7.3 MFA CHALLENGES

As previously said, multi-factor validation (MFA) is a workable security measure, but it is not without challenges and risks. This segment centers on a few of the key challenges and considerations related to MFA [11]:

e. The User Resistance: Clients may discover MFA forms awkward, driving to resistance in embracing the innovation. Additional steps that require inputting codes or utilizing biometrics ought to be obvious as being improperly organized.

f. Usability: The usability of MFA systems will change. Some systems, such as SMS-based codes, can be more difficult to understand than others, which can lead to errors and inconvenience.

g. Implementation Expense: Maintaining and configuring an MFA system can be quite expensive. Expenses may include equipment tokens, program progress, assigning tasks, and increasing maintenance.

h. Integration Complexity: Integrating MFA into existing systems can be time-consuming and complex. Heritage properties can have similar problems and require a lot of work and assistance.

i. Backup Tools: Organizations need to have backup tools in place to ensure customers can access their secure data in the event of an MFA failure or customer failure. Adding MFA is not enough.

j. Phishing Attacks: While MFA doesn't necessarily prevent phishing attempts, it can protect against many threats. If the user encounters an incident, the attacker can trick the user into providing an MFA code or token.

k. Lost or Deleted Device: Customers may encounter problems if their MFA device is lost or stolen. The organization must have procedures in place to deal with lost documents or stolen.

7.4 MFA RULES AND REGULATIONS

Multi-factor authentication is essential to ensure the security and functionality of MFA applications across multiple systems and services. Organizations can use MFA methods and standards to ensure the stability, consistency, and interoperability of the accreditation process [12]. Following standards when using MFA can increase security and provide a better user experience [13].

7.5 APPLICATIONS OF MFA

Two-factor authentication increases security by requiring more than one type of authentication before logging in, making it suitable for many businesses and companies. The use of MFA continues to grow as businesses and individuals realize the importance of improving security in a digital and continuous world [14]. MFA is a solution for data protection, privacy, and access to critical services and systems. This text covers important MFA practices, including a brief discussion of each [15] (Table 7.1).

7.6 INTELLIGENT SYSTEMS AND MFA

Cyberattacks pose a serious risk to operations and products related to robots, commercial IoT, artificial intelligence, computers, and other products used. This technology is an easy target for cybercriminals; ransomware attacks are the most common target on the internet.

Industrial Networks: Violations can halt production and operations for days or weeks, and these strikes often remain silent until they are extinguished. It is of the utmost importance to have different security technologies to prevent such attacks. This technology needs to be kept secret to prevent hackers from knowing the security measures used in smart machines. An effective measure is the implementation of MFA in the production environment. This not only improves security, but also ensures product availability and system performance.

Intelligent systems can use different types of MFA and some general concepts have been discussed in the previous sections. Two-factor authentication is frequently used and allows only people with the correct credentials to enter the system environment and access applications in the factory. In addition, authorization procedures at various levels, called micro-authorization, are used at various stages of production. These micro-authorizations provide an additional layer of security by allowing administrators to perform additional authorizations at runtime. These requests are necessary to keep security levels high and to deliver a positive, smooth user encounter.

7.7 TWO-FACTOR VALIDATION

This type of MFA strengthens the security of an exquisitely designed system by requiring customers to complete the two stages, also known as the two sections, in order to confirm their identities and arrange to access the system. Another security factor that comes to light is the username and secret word that the client just recently realized they entered the system. This information may be seen on a login security application that appears on their phone or device. The system of creation requires all the affirmation arrangement as well as the specific information of the customers.

By requiring customers to complete the two stages, also referred to as the two regions, to validate their personalities and arrange to access the framework, this type of MFA strengthens the security of a well-thought-out framework. The username

TABLE 7.1
Multi-Factor Authentication Applications

MFA Applications	Description
Websites and Online Accounts	• MFA is frequently used in customer accounts on websites and online services to protect customer data. • This covers financial institutions, email services, social networking platforms, and e-commerce websites.
Business Safety	• MFA is widely used by companies to safeguard critical corporate information and systems. • MFA is usually used by staff members to gain access to cloud-based services, VPNs, and workplace networks.
Distance-Based Access	• MFA is necessary for safe remote access to company networks, particularly for workers who are mobile or work from home. • MFA is frequently used by remote desktop services and virtual private networks (VPNs).
Exchanges of Money	• In the financial industry, MFA is frequently used to safeguard stock trading platforms, mobile payment apps, and online banking.
Medical Care	• MFA is used by healthcare businesses to safeguard patient data, electronic health records (EHRs), and apps connected to the industry.
Public Services and Government	• MFA is used by public services and government organizations to provide secure access to benefits systems, tax filing, and government websites.
Education	• MFA is used by educational institutions to protect administrative systems, online learning environments, and student records.
Cloud Services	• MFA solutions are frequently provided by cloud service providers to improve the security of data, infrastructure, and applications hosted on the cloud.
Email Security	• Email accounts can be secured with MFA, shielding private correspondence and preventing unwanted access.
Social Media	• MFA solutions are offered by a multitide of social networking sites in order to improve account security, stop illegal access, and safeguard user information.
Critical Infrastructure	• MFA is essential for protecting vital infrastructure systems, including transportation networks, water treatment facilities, and electricity grids.
E-commerce	• MFA is frequently used by online merchants to safeguard payment information and client accounts.
Travel and Hospitality	• MFA is used by hotels, airlines, and travel firms to safeguard client reservations and reward program accounts.
IoT Devices	• To prevent unwanted access to smart home appliances and industrial IoT systems, MFA can be incorporated into the administration of Internet of Things (IoT) devices.
Secure File Sharing	• Files transferred via cloud-based file storage and collaboration systems are secured with MFA.
Telecommunications	• MFA is a security measure used by telecom providers to protect consumer accounts, especially while accessing account details or making changes to accounts.

TABLE 7.1 (CONTINUED)

Multi-Factor Authentication Applications

MFA Applications	Description
Legal Services	• MFA is used by law firms and lawyers to protect client data, case files, and correspondence.
E-government Services	• MFA is frequently used by government websites to secure user data while providing services like tax filing, voter registration, and license renewal.
Identity Verification	• MFA is used for identity verification in a number of situations, such as user registration and account recovery.
Passwordless Authentication	• Password less authentication techniques, which do away with the need for conventional passwords and increase simplicity and security, are a growing area of use for MFA.

and secret word that the customer just recently began to comprehend when they joined the framework were another security aspect that were revealed. This information may appear on a security application that requires a login on their phone or device. This amount is anticipated to combine the demands of the customers and the certification arrangement into the creation of a safe environment of customers.

As a result, the two-factor authentication handle provides an additional push for the developers to take a break before entering the structure, and it provides an additional push for the programmers to halt or anticipate the software engineers' section. As a result, the customer may recognize proof with a higher degree of safety, and the record is anticipated in a few cases. These days, the trend is shifting as more and more smart systems use Fast Identity Online (FIDO)-based checks, which might be implemented across the association. This adds an additional degree of security since passwords have the potential to be hacked periodically, making them a further line of defense against system access. For an additional degree of protection, the client must know a mystery word, a combination of device tokens, or a security token that may be utilized and will be difficult for software programmers to decipher. Thus, the confidential data may be protected from clients who duplicate it in this fashion. By using 2FA security features, individuals and organizations lower their exposure to phishing attempts, weak passwords, and accreditation theft, protecting their clients' modernized device characters.

There are a few challenges that come with implementing 2FA in excellent systems, such as large values, laborious implantation, and similarity problems with intricate structures. UASB Client Get to Affirmation Representative is the PC application that serves as a blueprint for all of these joining problems. It streamlines the 2FA allocations, making the entire process ready, logical, and practical for any astute system where it is to be finished.

The technology allows security executives to download and employ 2FA into any online service without requiring the advancement of a PC software. This plan preparation is very quick since it just takes a few minutes and it can be completed with almost little planning. Convincing others to follow, verifying, learning the action plans, and switching between intermediary movement learning tools are some of the strategies that can assist in implementing 2FA tools in a creatively organized manner. Specific associations are locked in to overcome all the challenges associated with the 2FA options.

Therefore, it can be concluded that UASB is a powerful and highly configurable device that facilitates a broad range of applications for two-step verification processes. In addition, UASB provides a flexible environment for promoting the 2FA approach; this is why it is regarded as a freethinker plan. The 2FA part's flexibility is intended to allow it to be customized, altered, managed, and coordinated to satisfy the needs of the customer.

FIDO 2, an open online authentication application standard, is also included in the display. It is renowned for its improved phishing defenses and customer comfort, and because of its better architecture, it also strengthens the components of the conventional 2FA system. When it comes to clients with particular needs related to the 2FA components, FIDO 2 affects how standard procedures that rely on one-time client requests—like SMS confirmation or TOTP verification through various applications—are carried out. Associations choose the most appropriate 2FA component based on what they believe best fits security objectives, requirements, and customer preferences. This ensures a better, more comprehensive, adaptable, and secure authentication process.

7.8 CONCLUSION

It may be argued that later innovative forms of advancement and clever systems play a major role in the mechanical change. MFA plays a fundamental role in this. As the mechanically intricate structures advance, so do the security risks they raise. To address these concerns, an appropriate and proactive strategy must be put in place. Different types of safety difficulties are evident at obvious levels of the plans, and as a result, changes are anticipated to the security courses of action. Since verification serves as the initial point of entry for many problems and hazards, confirmation security concerns provide a strong explanation for the obvious security problems that appear at several levels. Consequently, the best plan of action to address the affirmation concerns in the excellent systems is to perform many workout checks. This section evaluates a variety of MFA types and their applications, including how they could be utilized in systems that make sense. With 2FA and 3FA, redesigned security plans are provided, adding an additional layer of protection to everything within the smart structures.

REFERENCES

1. T. Pereiraa, L. Barretoa and A. Amarala, Network and information security challenges within industry 4.0, *Manufacturing Engineering Society International Conference, Science Direct*, 13 (2017), 1253–1260.

2. Ahmad Barari, Marcos de Sales Guerra Tsuzuki, Yuval Cohen and Marco Macchi, Editorial: intelligent manufacturing systems towards industry 4.0 era, *Journal of Intelligent Manufacturing*, 32 (2023), 1793–1796.

3. Zohaib Jan, Farhad Ahamed, Wolfgang Mayer, Niki Patel, Georg Grossmann, Markus Stumptner and Ana Kuusk, Artificial intelligence for industry 4.0: systematic review of applications, challenges, and opportunities, *Expert Systems and Applications*, (2023), 216. doi: 10.1016/j.eswa.2022.119456

4. Amit Kumar Tyagi, Terrance Frederick Fernandez, Shashvi Mishra and Shabnam Kumar, Intelligent automation systems at the core of industry 4.0, *Intelligent Systems Design and Applications*, (2020), 1–18. https://doi.org/10.1007/978-3-030-71187-0_1

5. A. Ometov, S. Bezzateev, N. Mäkitalo, S. Andreev, T. Mikkonen and Y. Koucheryavy, Multi-factor authentication: A survey, Cryptography, 2(1) (2018), 1.

6. K. Abhishek, S. Roshan, P. Kumar and R. Ranjan, "A comprehensive study on multifactor authentication schemes, In *Advances in Computing and Information Technology: Proceedings of the Second International Conference on Advances in Computing and Information Technology (ACITY)*, July 13–15, 2012, Chennai, India. (pp. 561–568). Springer Berlin Heidelberg.

7. A. Ometov, V. Petrov, S. Bezzateev, S. Andreev, Y. Koucheryavy and M. Gerla, Challenges of multi-factor authentication for securing advanced IoT applications, *IEEE Network*, 33(2), (2019), 82–88.

8. C. Jacomme and S. Kremer, An extensive formal analysis of multi-factor authentication protocols, *ACM Transactions on Privacy and Security (TOPS)*, 24(2) (2021), 1–34.

9. M. Bartłomiejczyk and M. Kurkowski, Multifactor authentication protocol in a mobile environment, *IEEE Access*, 7 (2019), 157185–157199.

10. M. A. Ali, B. Balamurugan, R. K. Dhanaraj and V. Sharma, IoT and Blockchain based Smart Agriculture Monitoring and Intelligence Security System, *2022 3rd International Conference on Computation, Automation and Knowledge Management (ICCAKM)*, Dubai, United Arab Emirates, 2022, pp. 1–7, doi: 10.1109/ICCAKM54721.2022.9990243.

11. V. Sharma, N. Mishra, V. Kukreja, A. Alkhayyat and A. A. Elngar, Framework for Evaluating Ethics in AI, *2023 International Conference on Innovative Data Communication Technologies and Application (ICIDCA)*, Uttarakhand, India, 2023, pp. 307–312, doi: 10.1109/ICIDCA56705.2023.10099747

12. A. Raj, V. Sharma and A. K. Shanu, Comparative Analysis of Security and Privacy Technique for Federated Learning in IOT Based Devices, *2022 3rd International Conference on Computation, Automation and Knowledge Management (ICCAKM)*, Dubai, United Arab Emirates, 2022, pp. 1–5, doi: 10.1109/ICCAKM54721.2022.9990152

13. D. Preuveneers, S. Joos and W. Joosen, "AuthGuide: Analyzing Security, Privacy and Usability Trade-Offs in Multi-factor Authentication. In Trust, Privacy and Security in Digital Business: 18th International Conference, TrustBus 2021, Virtual Event, September 27–30, 2021, Proceedings 18. (pp. 155–170). Springer International Publishing.

14. A. Raj, V. Sharma, S. Rani, A. K. Shanu, A. Alkhayyat and R. D. Singh, Modern Farming Using IoT-Enabled Sensors For The Improvement Of Crop Selection, *2023 4th International Conference on Intelligent Engineering and Management (ICIEM)*, London, United Kingdom, 2023, pp. 1–7, doi: 10.1109/ICIEM59379.2023.10167225

15. C. Mohanraj et al., Conspiracy in the Stealing of Electricity Detection Through the IoT, *2023 3rd International Conference on Innovative Practices in Technology and Management (ICIPTM)*, Uttar Pradesh, India, 2023, pp. 1–5, doi: 10.1109/ICIPTM57143.2023.10117849

8 Lightweight Algorithms for Enhanced Privacy and Encryption Techniques for IoT-Based Intelligent Manufacturing

Aparna Sharma and Mahima Shanker Pandey

8.1 INTRODUCTION

The goal of the Internet of Things (IoT) is to improve our quality of life by giving us access to constant data that lets us keep an eye on our homes and workplaces from a distance. The Internet of Things (IoT) is a network of embedded devices that are linked to particular items to enable remote monitoring and reaction. IoT [1] can be applied in many different contexts, such as areas that are often not Internet-connected, such as food and garden, water and wastewater, front lines (i.e., the Internet of Military and Battle Zone Things, or IoT), and even dams. The utilization of IoT in the modern area era is referred to as the Modern Industrial Internet of Things (IIoT), which is additionally an abbreviation. Nonetheless, security is a significant concern for IoT since many conventional security measures might not be appropriate for the gadgets that comprise the network. In addition furthermore, security features of IoT- enabled devices are frequently ignored, which leaves them open to hackers. Recent research has found that many Internet of Things IoT devices are susceptible to attack because they have insecure default security settings, passwords that are easy to guess, and unencrypted network interactions.

Attackers are focusing more and more of their attention on IoT devices in an effort to purloin personal information and execute Distributed Denial of Service attacks. For example, the massive Mirai botnet attack of October 2016 nearly brought the Internet to a complete halt by exploiting vulnerabilities in smart devices. The Mirai botnet is a piece of software that uses the telnet protocol to look for devices—such as DVRs, security cameras, and smart home routers—that have their credentials set to default. Nonetheless, a number of Mirai versions that are capable of infecting a wide variety of devices have been discovered in the past two years. Additionally, security lapses in the chemical, oil and gas, utility, and industrial manufacturing industries can result in severe bodily harm, environmental catastrophes, and widespread

DOI: 10.1201/9781032630748-8

contamination. In other words, attacks on the Internet of Things have the capability to do enormous damage.

The Internet of Things is currently receiving a lot of attention because of the amazing prospects it presents for a variety of cutting-edge programming projects. From a broader angle, IoT proposes organising physical objects—such as vehicles, home appliances, cell phones, and many more—that link to and share data with personal computers. In an effort to capitalise on the numerous economic prospects presented by IoT, numerous previously established and recently established businesses have started to concentrate on this innovative field. It has a significant effect on a number of areas, including business process management, horticulture, tracking, supply chain management, continuous financial analysis, and other relevant fields.

The more private information such devices generate, the greater the risk of fraud, extortion, and information breaches. The lives of individuals and the operations of enterprises could be severely harmed by even little errors in the system if people grow unduly dependent on this cutting-edge innovation and automation. This could happen on both a practical and financial level. It also calls into question the usage of digital weapons. This is why it has attracted the attention of academics and scientists from all over the world.

With regard to the Internet of Things, Thabit et al. [2] allude to one of the most common ways for ensuring the wellbeing of IoT parts like gadgets, organisations, information, working frameworks, and servers. It is challenging to confirm the security of individual facts and data gathered by IoT gadgets because of the way that billions of gadgets are associated with each other through the Web. In case of a security break, having an alternate course of action prepared to execute to keep however much information as could be expected safe is totally fundamental.

Unauthorized real access to IoT devices can cause a catastrophic system failure in any setup, despite its benefits. Programmers can take advantage of the weaknesses in the system from a distance, negating the need for them to physically access the IoT device. Furthermore, an attacker may be able to take advantage of weaknesses in the framework code by breaching the operational structure of an Internet of Things group, which would give them complete control.

The IoT is a technical innovation which seeks to make our lives easier by giving us access to constant data that lets us keep an eye on our homes and workplaces from a distance. The IoT is a network that uses implanted devices connected to ubiquitous objects to enable remote observation and response. The IoT has a wide assortment of utilisations, some of which incorporate areas that are not commonly associated with the Web, such as dams, food and farming, water and wastewater frameworks, and perilous conditions as combat zones (otherwise called the Web of Military and Front Line Things). The use of IoT in the modern era is referred to as the Modern Internet of Things (IIoT), which is likewise an abbreviation. Nonetheless, security is a significant concern for the Internet of Things since many conventional security measures might not be appropriate for the gadgets that comprise the network. Furthermore, IoT-enabled devices' security measures are routinely disregarded, leaving them vulnerable to hackers. According to recent research, a lot of Internet of Things devices have weak default security settings, easily guessed passwords, and unencrypted network connections, which make them vulnerable to attack.

Attackers are concentrating more and more of their efforts on IoT devices to commit Distributed Denial of Service (DDoS) attacks and steal personal data. For instance, by taking advantage of weak points in smart devices, the extensive Mirai botnet attack that was done in October 2016 almost brought the servers to a total stop. The software known as the Mirai botnet searches for devices, including security cameras, DVRs, and smart home routers that have their credentials set to default values via the telnet protocol. However, in the last two years, a number of Mirai variants that can infect a broad range of devices have been found.

The public sector presently enjoys several types of benefits from the IoT, and in the near future, the public could profit greatly more from it. However, creating IoT devices without providing sufficient privacy protections for consumers may result in unfavourable outcomes. Large volumes of data are produced as an outcome of the expansion of IoT devices. Privacy concerns have been raised since it is possible that these data will contain private, sensitive health information. Smart energy metres, for example, have the ability to reveal a range of personal data about people and how they use the appliances in their homes.

The main roles of assault relief are to shield the protection and classification of clients, secure the security of Internet of Things foundations, information, and gadgets, and assure the accessibility of administrations given by an IoT biological system. Since the data that these gadgets gather will be used to go with significant decisions, they should stay on the Web and capability precisely consistently for IoT applications to fill in as expected. For us to have the option to have total confidence in the dynamic cycle, one of the difficult errands that must be finished is to confirm the validity of the gadgets. Notwithstanding the manner in which the executives are trusted is turning out to be progressively famous because of its capacity to stop or distinguish maverick hubs, validation is, as yet, the security procedure that is used by far most of the time. Then again, the essential focal point of encryption research is on cultivating minimal expense and low-weight encryption strategies that are appropriate for low-power and obliged gadgets.

The term "Modern Web of Things" refers to the use of intelligent actuators and sensors to improve manufacturing and other contemporary cycles. The concept of the Modern Web of Things, also referred to as Industry 4.0 or simply the modern Web, makes use of ingenious devices and ongoing analysis to leverage the data that "moronic machines" have been providing in contemporary environments for an extended period of time.

There are many measurable benefits of IoT for manufacturing. Therefore, there is a need for more researchers to use the Industrial Internet of Things IIoT so that industrial processes can be improved. IoT and IIoT are expected to have 75.44 billion deployed devices by 2025, and by 2030, the IIoT is expected to boost the world economy by US$14.22 trillion, according to Accenture.

8.1.1 Evolution of Internet of Things

A highly interconnected world known as the IoT is being created by a combination of machine-to-machine, or M2M, communications and connected device exchanges. It has exabytes of data generated by sensors, a large number of exabytes and a large

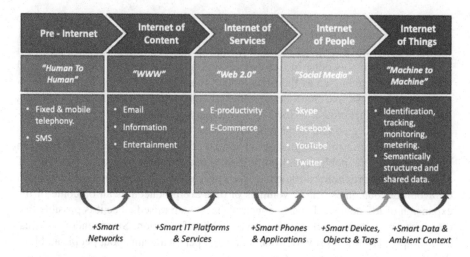

FIGURE 8.1 The IoT evolution.

number of intelligent devices in addition to strong rates of economic growth and cost reductions. It is generally seen as the potentially revolutionary next phase in the evolution of the Internet. The Internet was thought to be nothing more than a static repository of interconnected hypertext documents a few decades ago. However, it has developed into a robust network of people, devices, and applications that collaborate in the present period. As of late, all that has been designed starting from the earliest stage for Web availability, including cars, refrigerators, clothes washers, watches, apparel, and health monitoring frameworks. The Web of Administrations has supplanted the Web of Content as its true name (Figure 8.1).

8.1.1.1 The Internet of Content

The first Web was launched in the middle of the 1990s thanks to the Hyper Text Move Convention (HTTP) and the Internet (www). This marked the beginning of what is today called the Web of Content. By then, the Web had essentially become static and was used for sharing and publishing content.

8.1.1.2 Internet of Services

User-generated content has been made possible on the Internet in more recent times with the backing of Web services, XML, and a wide range of commercial activities. With the introduction of Web 2.0, websites may now post dynamic pages instead of the more static pages that were typical of early websites due to advancements in productivity and collaboration tools.

8.1.1.3 Internet of People

The widespread use of smartphones and tablets, the accessibility of reasonably priced mobile broadband connectivity, and the rising acceptance of social networking software have all contributed to the development of the Internet's third stage.

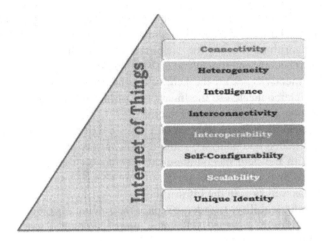

FIGURE 8.2 Characteristics of Internet of Things.

8.1.1.4 Internet of Things

This marks the start of the next revolution in how people use the Internet—a shift that will ultimately lead to a revolution. This revolution is made possible by the analysis of vast volumes of data and the networking of machines with one another. Everyday life is transformed as objects are interconnected and integrated into information systems and end-user apps. They create a ubiquitous, infinite, and networked universe where humans and robots work together to create a safer, healthier, and more ecologically responsible civilisation [3].

8.1.4.1.1 Characteristics of Internet of Things

Because it incorporates the convergence of multiple different disciplines, "Internet of Things" can be thought of as an umbrella word. The data that is assembled by the IoT gadgets, sensors, and actuators that are introduced in the actual climate adds to the upgrade of the worth of the organisation (Figure 8.2).

The following is a list of the primary properties of the IoT:

1. **Connectivity:** Connectivity is important because it improves both the accessibility and compatibility of a network. Compatibility refers to the shared capacity to utilise and produce data, whereas accessibility refers to the process of gaining access to a network.
2. **Heterogeneity:** The nature of the devices connected to the Internet of Things is heterogeneous due to the fact that they employ a variety of hardware platforms and are connected to a variety of networks. These devices are able to communicate with other devices or platforms by utilising a variety of network protocols.
3. **Intelligence:** In the Internet of Things, intelligent things are interconnected and equipped with appropriate interfaces in the digital realm for the

purpose of communicating and computing the data arriving from a variety of sources. The use of sophisticated data mining and knowledge representation algorithms provide the "intelligent spark" that enables inanimate objects to become intelligent and assists them in making decisions.

4. **Interconnectivity:** With the help of the Internet, anything and anybody may be interconnected to the Internet of Things where they can exchange information and speak with one another.

5. **Interoperability:** In an Internet of Things environment, vast amounts of data are traded and analysed using a wide variety of models and file formats. With the help of communication protocols that are interoperable, the linked devices that make up the IoT are able to communicate with one another as well as with the infrastructure.

6. **Self-Configurability:** In order to reduce the amount of human involvement that is required, the devices and items comprising the IoT must possess the capability to react autonomously to a wide range of situations. Therefore, these intelligent gadgets have the capacity to configure themselves with IoT infrastructure, set up networking, and retrieve the most recent software upgrades with less intervention from the user. They are able to organise themselves independently with the networks in order to provide the fundamental medium and carry out coordinated functions. One example of this is their capacity to undertake device and service discovery without the need for an external trigger.

7. **Scalability:** The number of devices and objects which can be connected to an IoT ecosystem is growing at an exponential rate today. According to estimates provided by Gartner, the IoT connects approximately 5.5 million new items on a daily basis, and the total number of linked devices is expected to reach 20.8 billion by the year 2020.

8.2 CONVENTIONAL IOT ARCHITECTURE

Three layers make up this specific design: the application layer, the network layer, and the perception layer [4] (Figure 8.3).

The three layers listed below explain the fundamental concept behind the Internet of Things.

i. **The perception layer:** This layer, also known as the recognition layer, is the same as the physical layer. Throughout the natural environment, sensors and actuators are used at different places to perceive and gather data. Sensors may detect the physical elements of their environment or identify other sentient things in their immediate area. This component ensures the secure transfer of data between the application and perception layers by acting as the brains of typical Internet of Things architecture. This layer manages the convergence of communication-based networks and the Internet.

ii. **The network layer:** Connectivity between servers, smart devices, and other network equipment is established via this layer. It is responsible for

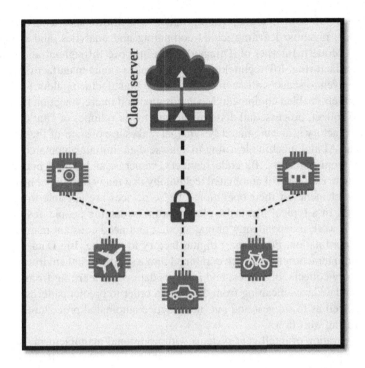

FIGURE 8.3   A Conventional IoT architecture.

sending and analysing the data that the sensors have gathered. The layer is also in charge of data processing and administration and controls the various addressing and routing functionalities. This layer contains contributions from the satellite, wireless, and wired technology domains.

iii. **The application layer:** In a typical Internet of Things design, this is the top layer that bridges the gap between consumers and apps. The tasks of this layer are limited to providing application-specific services to the user. It addresses the use cases for the IoT, including smart cities, smart homes, and smart health.

8.3   WHAT IS INTELLIGENT MANUFACTURING?

Manufacturing that uses a range of technology integrations, such as automation, data collecting, artificial intelligence, machine learning, and others, to produce ideal production circumstances is referred to as "intelligent manufacturing," also known as "smart manufacturing." It is frequently used interchangeably with Industry 4.0, or the Fourth Industrial Revolution, to refer to the shift toward digital production and increased connectivity.

The way that businesses manufacture, optimise, and advertise their goods is being profoundly impacted by Industry 4.0. Manufacturers are incorporating new technologies into their operations and production facilities, such as artificial intelligence (AI), machine learning, cloud computing and analytics, and the Internet of Things. Modern Internet of Things devices improve throughput and quality in smart manufacturing. IoT technology is important to smart manufacturing. The IP addresses given to sensors allow machines on the manufacturing floor to share data with other web-enabled equipment. Mechanisation and interconnection have made it feasible to collect, process, and disseminate enormous volumes of vital data.

A smart factory is an automated system that makes use of state-of-the-art technology such as AI and machine learning to analyse data, initiate automated processes, and learn from its errors. Barcode scanners, cameras, and digital manufacturing equipment are examples of automated technology that many traditional firms employ for various elements of their operations. These devices are not able to speak with one another. In a typical plant, the platforms for managing people, resources, and information work independently of one another and need constant manual integration and coordination. For a smart digital factory to operate, Big Data, equipment, and human interaction must all be combined into a single digital environment [5]. A smart factory collects, organises, and assesses data while learning from the past. It analyses and deduces meaning from datasets in order to predict patterns and occurrences, as well as to suggest and put into practice automated procedures and smart manufacturing workflows.

The integration of intelligent systems with traditional manufacturing sectors has been made possible by the advancement of enabling technologies. Radio Frequency Identification (RFID) technology allows for the automatic identification of hard resources. Wireless sensor networks (WSN) possess the ability to recognise, gather, and interpret critical real-time data produced by manufacturing process machines. Wireless communication technology allows for the efficient transmission of data to various production elements of the system, including RFID and WSN [6,7].

The Internet of Things is a subset of the Industrial Internet of Things (IIoT), which highlights its uses in modern manufacturing and industries. As such, it is ideally suited to the construction of intelligent manufacturing firms. The primary drivers of the IIoT market's growth include also the availability of automation tools, increased data rates, and communication technology reach, expanding cloud computing platform utilisation and the rising adoption of Internet Protocol version 6 (IPV6). When combined with the right services, networking technologies, apps, sensors, software, middleware, and storage systems, IIoT also offers features that make it possible to gain insight and manage business processes and assets [1].

Manufacturing is considered a fundamental industry, impacting people's livelihoods and a country's economy significantly. It covers a wide range of activities, procedures, services, and products and is one of the biggest and most closely related IoT sectors [1].

8.4   INTERNET OF THINGS AND INTELLIGENT MANUFACTURING

A technological revolution has been sparked by the concept of the IoT, which made it possible for physical things and digital networks to cohabit together. Commonplace objects, business equipment, and even automobiles will be equipped with sensors, connections, and intelligence to collect and exchange data among themselves in the era of IoT. Due to the shift in paradigm, a lively network made up of a collection of interlinked devices has existed, which is called the Internet of Things. This has made it possible to maintain links beyond the human-computer interaction.

Smart manufacturing will not be possible unless the digital revolution replaces traditional workplaces with automation and advanced technologies that make work more efficient and easier. This change is designed to transform the way business is done, enabling decision-making, automation, continuous monitoring and predictive maintenance through smart manufacturing based on the Internet of Things. Businesses can now increase productivity, reduce costs, reduce operating costs, and increase efficiency. IoT is the foundation of intelligent design based on the concept of the Internet of Everything.network; it allows the exchange of information, ideas, and concepts about materials, production lines, machines and entire factories. This improves the flow of information. This discussion environment encourages collaboration between previously independent businesses, encouraging the creation of business models that can transform business models and business expectations. Network communication and data exchange have created significant problems, especially in terms of security and privacy. The increase in the number of devices connected to the network increases the risk of cyber threats and attacks. This vulnerability opens the door to unauthorised access, deletion of data, and unauthorised surveillance. These attacks can disrupt operations, destroy productivity, and expose confidential or personal information. Data protection through encryption and privacy measures should be emphasised.

8.4.1 IoT in Manufacturing: Transforming Industries

The Internet of Things is impacting many industries, including manufacturing. The IoT has transformed factories, making them smarter and beyond traditional operations. It is changing the design landscape by connecting systems, devices, and operations, enabling data-driven analytics and maximising value through digital connectivity. In today's world, production is more than just a physical activity. It is the product of a complex process of creating and analysing data. Thanks to the IoT, sensors built into devices can collect new data such as temperature, humidity, pressure, and activity. This information is transferred and processed quickly to support accurate repair planning. These programs help prevent equipment failure, reduce downtime, and increase productivity through equipment maintenance and inspection. IoT has revolutionised supply chain management by providing a comprehensive view of all stages of production. Companies can track raw materials, production, and delivery on time using technologies such as RFID tags, GPS tracking and stock management. This transparent development reduces waste and increases efficiency while also promoting environmentally friendly practices in materials. IoT enables the development of cyber-physical systems by bridging the gap between information technology (IT) and operational technology (OT). These systems connect digital data analysis and cloud computing to the physical world of machines. Thanks to instant messaging, companies can improve production processes, quickly respond to market changes and make data-driven decisions (Figure 8.4).

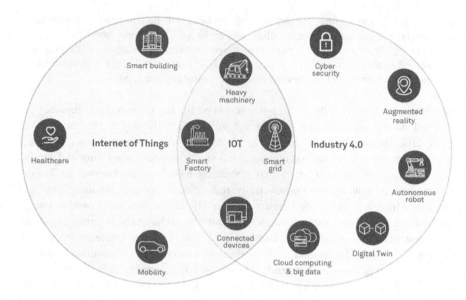

FIGURE 8.4   IOT in manufacturing industries.

8.5   APPLICATION OF IOT IN INTELLIGENT MANUFACTURING

8.5.1   IN QUALITY CONTROL

"Quality assurance" in the production process ensures that the product complies with standards. "Quality Control" involves verifying that products meet strict specifications and satisfy customers. In the past, quality control involved inspectors manually examining each product for defects and compatibility. Thanks to advances in IoT quality control, companies can use real-time product monitoring and data analysis to detect quality issues early. This allows for faster adjustments and improvements, improving quality and customer satisfaction. Thermal and video sensors help IoT improve the field of quality control by collecting a wealth of data about products at different stages of production. Developers can monitor systems in real time to detect errors before they occur and use IoT devices to collect the data needed to fix problems by viewing control predictions.

8.5.2   IN INVENTORY MANAGEMENT

The Internet of Things uses radio frequency identification (RFID) to make inventory management faster and easier. Each unique identification number (UID) associated with a product in the stockpile is derived from the digitally encoded data on the RFID tag. The tags may be scanned by RFID scanners, and once the data is collected, it is forwarded to the cloud for processing.

Industrial IoT makes it easier to convert the data collected by RFID scanners into insightful business information. It provides users with comparison results by tracking supply chain activities, stock product locations, and status. For instance, an IoT-driven inventory management system may calculate the quantity of supplies needed for a future production cycle based on data on inventory location and quantity. The system can notify users when to restock supplies and when a certain inventory item goes.

8.5.3   In Forecast Maintenance

Time-dependent forecasting has historically been used by manufacturers to schedule maintenance for their machines and tools. A time-based approach is unproductive and may prove to be costly over time. Manufacturers can avoid such inefficient maintenance procedures by utilizing data science with industrial IoT for predictive maintenance. The equipment's IoT sensors allow them to keep a close watch on its working conditions and conduct analytics utilising relevant cloud data to assess the real wear and tear. Fast service and repair provide greater cost savings as well as improved field technician job allocation, downtime avoidance, and maintenance process efficiency.

8.5.4   In Operation Safety

IoT also improves a manufacturing plant's operational, equipment, and worker safety. Indicators such as employee absences, collisions with vehicles, equipment damage, and any other incidents that interfere with regular operations can be monitored using it. When working in factories and fields, employees can also have their health parameters continuously monitored by using wearables connected to the Internet.

8.5.5   In Smart Metering

The manufacturing, utility, and other industries have been able to access the world of smart meters thanks to the Internet of Things. These smart meters monitor fuel, water, and energy usage, enabling companies to evaluate individual usage and put policies in place for more effective resource management. Smart grids depend on the data these smart meters gather since it gives utility companies the ability to monitor and regulate service delivery flows remotely and thoroughly. Through smart planning, utilities can monitor and achieve goals related to climate change and environmental impacts while providing plans that include sustainability and good and safe performance.

8.5.6   Smart Packaging

Smart packaging integrates data connections to enable manufacturers to leverage the full potential of the Internet of Things. It allows instant communication with customers and provides useful information to increase consumption. Creative packaging often includes videos, beauty tips, recipes and other content. Different IoT technologies, including sensors, QR codes, and reality (augmented, virtual, and hybrid), interact with packaging in different ways.

8.5.7 DIGITAL TWINS

The Internet of Things can create "digital twins", which are representations of objects or systems. These digital twins provide a deep understanding of business processes to support design, analysis, and optimisation.

8.6 PRIVACY AND SECURITY CONCERNS IN IOT-BASED INTELLIGENT MANUFACTURING

Through smart strategic planning, utilities can monitor and achieve goals related to climate change and environmental impacts while fulfilling their planning responsibilities. Powered by IoT, intelligent design is transforming the business environment through connected devices and data-driven processes. However, this change also raises security and privacy concerns. Much of the data generated by IoT devices, including data that measures performance and innovation, raises creativity and privacy issues. It is important to establish ownership of the crystal and monitor the use of the material to prevent illegal use. Additionally, connectivity further exposes production processes to vulnerabilities such as data leakage, unauthorised access, and cyberattacks. Implementing effective access controls, encryption methods, and detection tools is crucial to maintaining data integrity and preventing data breaches. The lack of security measures for IoT devices in production raises additional security concerns. Changes in security and communications will create vulnerabilities that hackers can exploit. Additionally, connected devices are easier to attack because they can compromise the security of the entire system. When employee access to IoT data is restricted, the potential for insider threats increases, leading to unauthorised access or unauthorised access to data. Extensive training, effective access control, and regular maintenance are required to reduce these risks. Improving data transfer and storage security is very important for IoT systems. Data must be sent and stored securely to prevent unauthorised access and interception. Monitoring compliance is also important. The Internet of Things often contains sensitive data, so business and personal data need to be processed in accordance with standards. Meeting these standards, especially the EU General Data Protection Regulation, helps maintain legal compliance and stakeholder trust. IoT devices can collect personal data without the user's knowledge, raising privacy concerns. These issues require fair decision-making focused on anonymising data and using strong privacy protections. Integrating IoT with existing systems can also provide security improvements; therefore, these risks need to be addressed [11–15].

8.7 DATA PRIVACY IN INDUSTRIAL IOT

The Internet of Things industry connects machines, sensors, and devices to create comprehensive data. Data privacy is critical to protecting sensitive information created and shared in this environment. It includes ensuring compliance and confidentiality while maintaining data integrity. Personal information includes protected personally identifiable information (PII), confidential business information, personal information, and business processes. This should prevent access to data collected by sensors embedded in equipment, supply chains, and production lines.

Encryption is important to protect data on Industrial Internet of Things (IIoT) devices. It protects sensitive data stored and transmitted on these devices from unauthorised access. Only authorised users with the decryption key can access and decrypt files. This security protection prevents information from being compromised even during public meetings. Anonymised data is also used to protect privacy and facilitate data analysis. By removing personal identifiers, organisations can provide more information without revealing private or sensitive information. This supports collaboration with third parties and data transfer via various methods.

To ensure data privacy in IIoT, companies must comply with industry regulations and data protection regulations such as GDPR. This means obtaining explicit consent, giving people control over their data, and promptly reporting data breaches. Balancing the benefits of industrial IoT with data privacy requires a strategic approach. Organisations need cybersecurity measures such as vulnerability assessments, incident response plans, and regular monitoring. It is also important to create a culture that values privacy.

8.8   ENCRYPTION TECHNIQUES FOR SECURING IOT DATA

In the Internet of Things, encryption is essential to protect data security and privacy. As IoT devices and networks grow, data encryption must also increase. Encryption converts data into code that can only be decrypted by an authorised party with the correct decryption key. Encryption technology is critical to protecting sensitive data on devices and systems in the IoT environment.

1. **End-to-End encryption:** The basic idea is to provide protection from the moment the recipient leaves the website until he or she reaches the desired destination. This ensures that even if data is intercepted during transmission, it cannot be decrypted without the decryption key. Symmetric encryption uses the same key for encryption and decryption; this can improve performance but requires careful management of keys to prevent unauthorised access.

2. **Asymmetric encryption:** Asymmetric encryption, which is also known as public-key encryption, utilises a pair of keys: a public key for encryption and a private key for decryption. This approach enables secure communication between devices without the need of sharing a common secret key. However, it requires computational resources and is often used in combination with symmetric encryption for efficient data transmission. Homomorphic encryption enables data computations on encrypted data without the need to decrypt it first.

This method safeguards the privacy of data while processing, making it appropriate in situations where data analysis is required but security cannot be compromised.

Although encryption provides better security, it is important to manage keys securely, distribute keys efficiently and effectively, and store keys securely.

Additionally, IoT devices may be vulnerable to data attacks using encryption and decryption techniques. Measuring security through resource performance is particularly important in the IoT environment due to device capabilities and vulnerabilities.

In the IoT world, encryption is essential for data protection. It protects data shared between connected devices, ensuring it remains private and immutable. As the Internet of Things grows, it is important to choose encryption methods and key management systems carefully. These measures ensure that IoT systems are reliable and protect sensitive data from unauthorised access.

8.9 LIGHTWEIGHT ALGORITHMS TO IMPROVE PRIVACY AND IOT ENCRYPTION METHODS

Cryptography is a popular technique that uses mathematical formulas and legal techniques to protect information and communications. This is important to protect the confidentiality and integrity of your information. In the context of the Internet of Things, devices such as RFID and sensors have the ability and measure to create security according to their unique needs. IoT devices face issues such as insufficient memory, poor performance, short battery life, or inability to process RFID tags and physical locations.

Due to resource limitations, encryption may not be compatible with this device. Lightweight encryption provides solutions to these problems, including lower memory usage, reduced latency, improved performance and faster response times. Lightweight encryption technology limits IoT usage by ensuring secure communication and data protection on devices. In this case, using traditional encryption methods on IoT devices will not work.

On the other hand, lightweight encryption technology solves the shortcomings of traditional encryption methods by offering the advantages of low memory consumption, low requirements, power consumption, low power, and fast response time even on limited hardware.

8.10 DEVELOPMENT OF LIGHTWEIGHT CRYPTOGRAPHY

(i) A project to develop lightweight cryptography was started in Europe in 2004 and has recently been resurrected through the M2M/IoT process.

(ii) In the ISO/IEC JTC 1/SC 27 meeting, the internationally recognised standard ISO/IEC 29192 "Lightweight Cryptography" was created [9].

(iii) The Lightweight Cryptography Project was started in 2013 by the US National Institute of Standards and Technology (NIST), which publishes standards on cryptographic technologies, and in 2017 they released a call for applications for lightweight cryptographies.

8.10.1 TARGETED DEVICES

Targeting an extensive number of devices, lightweight cryptography can be implemented on a diverse range of hardware and software (Figure 8.5).

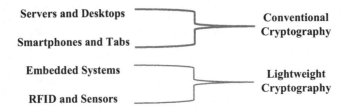

FIGURE 8.5   Architecture of proposed model.

Servers and desktop computers are at the high end of the device spectrum, followed by iPad and smartphones. These devices usually support conventional cryptographic algorithms well; as a result, these platforms do not need lightweight algorithms.

On the lower end of the spectrum are gadgets such as RFID devices, embedded systems, and sensor networks. This point in the spectrum is where the very restricted devices that are the main focus of lightweight cryptography are found. At the other end of the spectrum, RFID and sensor networks are frequently used in application-specific integrated circuits (ASICs) to meet some of the most stringent implementation requirements.

A tiny amount of power is provided by the environment to RFID tags which aren't battery powered. Such devices demand for methods of encryption that not only adhere to rigorous time and power limitations, but also employ relatively few gate equivalents (GEs).

An aggregator needs to make accommodations for the restricted sensors when they use lightweight cryptography approaches in order to communicate with them. To put it another way, the application and environment determine whether or not traditional norms are suitable. The need for lightweight cryptography stems from both the limitations of a single device and the other devices that an application directly interacts with.

8.10.2   Lightweight Cryptography

The focus of lightweight cryptography is on developing solutions especially for low-resource systems. The foundation for the creation of lightweight cryptography techniques is the AES (Advance Encryption Standard) algorithm, which is standardised by NIST. It is a symmetric block cypher that functions on 128-bit blocks with reversible key sizes of 128, 192, and 256 bits [8].

8.10.3   Categorisation of LWC Algorithms

8.10.3.1   Classification of LWC According to Structure

Symmetric key and asymmetric key are the two basic classifications for cryptographic algorithms (Figure 8.6):

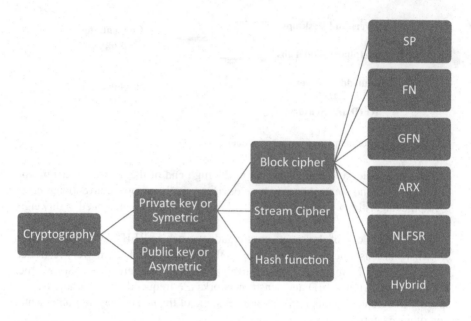

FIGURE 8.6 Classification of LWC according to structure.

(i) **Symmetric key** encryption and decryption only require a single key to encrypt and decrypt. The primary drawback of symmetric key cryptography is that it requires the communication parties to share the key without compromising it. Despite this, symmetric key cryptography is secure and relatively quick. But this may be avoided by giving the key to a reliable third party in advance. Additionally, it guarantees the data's security, accuracy, and authentication using authentication encryption mode.

(ii) **Asymmetric cipher** encrypts and decrypts data using two distinct keys. Two private-public key pairs are used in asymmetric cryptography. The receiver's public key is used to assure confidentiality and integrity, and the sender's private key (used as a digital signature to encrypt the data) is used to further ensure authentication. The recipient decrypts it at the other end by first utilizing the sender's public key and then his or her private key. The primary drawback of this type of encryption is its huge key, which makes the process extremely complex and time-consuming.

Block cypher is used in lightweight cryptography to generate cypher text. A block cypher generates an equivalent-sized block of ciphertext bits by utilising a block of plaintext bits. The block size in that specific system is fixed. The choice of block size has little bearing on the strength of the encryption technique. The strength of the cypher is affected by the key's length [16–18].

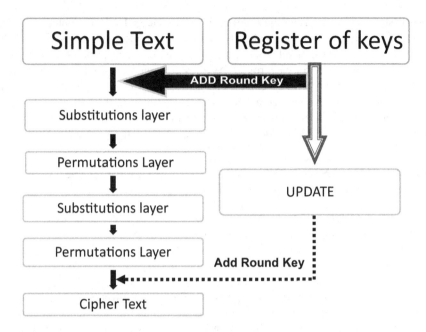

FIGURE 8.7   Description of SPN algorithm.

(iii) **Types of Block ciphers for LWC**
   a. **Substitution-Permutation Network (SPN):**

Substitution-permutation networks, or SPNs, are networks that take blocks of plaintext and keys and execute successive rounds of substitution layers (S-boxes) and permutations layers (P-boxes) to produce the final version of ciphertext (Figure 8.7).

First, the 64-bit plaintext is concatenated with the round key. The outcome is subsequently assessed by the Substitution layer. The round key used in ciphering is the 64 leftmost bits of a key register. The Key Schedule then modifies the 128-bit key register following every round. Only Present-80 is permitted to use the Key Schedule technique.

8.10.3.2   Feistel Structures

A general-purpose block encryption is the Feistel encryption. It functions as a template for design from which a variety of block cyphers can be created. The encryption method uses the Feistel structure, with a "substitution" step and a permutation step in each plaintext processing cycle.

The letters L and R stand for the left and right halves, respectively, of the input block for each round. R, the right half of the block, remains unchanged during each round. On the other hand, R and the encryption key determine what the left half, L, does. The encryption function "f," which accepts the keys K and R as inputs, is used first. f(R,K) is the function's output. L is then merged with the outcome of the mathematical function. In a practical Feistel Cypher implementation, such as DES, a round-dependent key, or subkey, is generated from the encryption key, rather than using the complete key for each round. As a result, even though all of these subkeys are connected to the main key, each round utilises a different key. The modified L and unmodified R are switched at the permutation phase at the end of each round [10]. As a result, the R from the current round would be the L for the following one. And the output L from the previous round serves as R for the subsequent one.

8.10.4 CHALLENGES AND SECURITY CONSIDERATIONS IN IOT-BASED MANUFACTURING

While IoT manufacturing has brought about revolutionary changes, it has also brought with it many security issue that need to be taken into consideration. As machines, sensors, and devices become more networked in this dynamic ecosystem, several important challenges arise in the technical, business, and security domains.

- **Publishing and data management:** Producing through the Internet of Things will create more complex data. Managing this and extracting important data from it can be a daunting task. Manufacturers need to invest heavily in data analysis and powerful systems to overcome these problems and not be influenced by the data.
- **Relationships and standards:** Companies face challenges in effectively collecting, managing, and interpreting important information. To prevent data loss, it is important for companies to invest in data management and analysis.
- **Scalability:** As IoT networks expand, capacity management becomes more difficult. It is important to create systems that can manage the number of interconnected devices without compromising security, performance, or speed.
- **Connectivity and reliability:** IoT devices need reliable connectivity to transmit and receive data efficiently. In the case of a cabling problem or cabling malfunction, operational problems and financial losses may occur. Managing effective communication can be difficult in many business environments.
- **Vulnerabilities:** IoT devices often face vulnerabilities due to insufficient memory and power consumption. Then malware attacks, unauthorised access, deletion of data, etc. There is a possibility, face many cybersecurity problems.
- **Privacy and compliance:** IoT devices often have limited resources such as memory and processing, making security difficult to manage. Because of this vulnerability, companies face cybersecurity risks such as malware infection, data theft, and unauthorised access.

- **Supply chain vulnerabilities:** The increasing use of the IoT in connected devices increases the potential for cyberattacks. This is because security conflicts in the chain become a target for hackers to access the chain, thus affecting the integrity of the chain.
- **Law integration:** Implementing IoT technology into the current business environment can be difficult, especially when incorporating legacy technology. Security and efficiency in connecting new technologies are important for protecting new and existing systems.
- **Insider risk:** Businesses connected using the IoT are vulnerable to (intentional and unintentional) security risks due to employees being able to access IoT devices. These risks may include illegal security breaches or misuse of customer information, as well as cyber threats.
- **Continuous monitoring and updating:** To protect IoT devices, vulnerabilities need to be monitored and remedied quickly. But maintaining consistent updates across multiple devices can be difficult.

8.11 CONCLUSION

This chapter examines privacy issues and security-related risks arising from smart manufacturing, specifically in the context of IoT devices. It proposes simple encryption and privacy management algorithms to make IoT security easier to use and operate. The analysis suggests ways to manage IoT risks from a manufacturing perspective, from addressing the technology to developing and implementing advanced security solutions While the Internet of Things and smart design have great benefits in terms of productivity, efficiency, and personalisation, the security and privacy issues they bring must also be addressed to ensure their security and sustainable development.

REFERENCES

1. Georgios Lampropoulos, Kerstin Siakas, Theofylaktos Anastasiadis. Internet of Things (IoT) in Industry: Contemporary Application Domains, Innovative Technologies and Intelligent Manufacturing, *International Journal of Advances in Scientific Research and Engineering*, 2018. https://doi.org/10.31695/ijasre.2018.32910
2. Fursan Thabit, Ozgu Can, Asia Othman Aljahdali, Ghaleb H. Al-Gaphari, Hoda A. Alkhzaimi. Cryptography Algorithms for Enhancing IoT Security, *Internet of Things*, 2023.
3. Usman Ahmad Usmani, Ari Happonen, Junzo Watada. Secure Integration of ioT-Enabled Sensors and Technologies: Engineering Applications for Humanitarian Impact, *2023 5th International Congress on HumanComputer Interaction, Optimization and Robotic Applications (HORA)*, 2023.
4. Uzair Javaid,Ang Kiang Siang,Muhammad Naveed Aman,Biplab Sikdar. Mitigating IoT Device based DDoS Attacks using Blockchain.
5. ElvisHozdić. Smart Factory for Industry 4.0: A review, *International Journal of Modern Manufacturing Technologies*, vol. VII, no. 1 / 2015.

6. Kamalakanta Muduli (Editor), V. P. Kommula (Editor), Devendra K. Yadav (Editor), M. Chithirai Pon Selvan (Editor), Jayakrishna Kandasamy (Editor). *Intelligent Manufacturing Management Systems: Operational Applications of Evolutionary Digital Technologies in Mechanical and Industrial Engineering*, Wiley, 2023.

7. Rajkumar Banoth, Rekha Regar. *Classical and Modern Cryptography for Beginners*, Springer Science and Business Media LLC, 2023.

8. Morteza SaberiKamarposhti, Amirabbas Ghorbani, Mehdi Yadollahi. *A comprehensive survey on image encryption: Taxonomy, challenges, and future directions*, Chaos, Solitons & Fractals, 2024

9. Kenneth J. Kepchar. A Systematic Approach to Influencing System Security Standards, *INCOSE International Symposium*, 2017.

10. *Computational Intelligence, Communications, and Business Analytics*, Springer Science and Business Media LLC, 2017

11. Yabing Chen, Fanglin An, Weili Jiang. ALICA: A Multi-S-Box Lightweight Cryptographic Algorithm Based on Generalized Feistel Structure

12. L. Dalmasso, F. Bruguier, P. Benoit and L. Torres, Evaluation of SPN-Based Lightweight Crypto-Ciphers, *IEEE Access*, vol. 7, pp. 10559–10567, 2019, doi: 10.1109/ACCESS.2018.2889790.

13. V. A. Thakor, M. A. Razzaque and M. R. A. Khandaker, Lightweight Cryptography Algorithms for Resource-Constrained IoT Devices: A Review, Comparison and Research Opportunities, *IEEE Access*, vol. 9, pp. 28177–28193, 2021, doi: 10.1109/ACCESS.2021.3052867.

14. Internet of Things in the Industrial Field. In: Pesado, P., Arroyo, M. (eds) *Computer Science – CACIC 2019. CACIC 2019. Communications in Computer and Information Science*, vol 1184. Springer, Cham. https://doi.org/10.1007/978-3-030-48325-8_22

15. V. Lesi, Z. Jakovljevic and M. Pajic, Security Analysis for Distributed IoT-Based Industrial Automation, *IEEE Transactions on Automation Science and Engineering*, vol. 19, no. 4, pp. 3093–3108, 2022, doi: 10.1109/TASE.2021.3106335.

16. Y. Huo, C. Meng, R. Li and T. Jing, An overview of privacy preserving schemes for industrial Internet of Things, *China Communications*, vol. 17, no. 10, pp. 1–18, 2020, doi: 10.23919/JCC.2020.10.001.

17. C. Yang, W. Shen and X. Wang, Applications of Internet of Things in manufacturing, *2016 IEEE 20th International Conference on Computer Supported Cooperative Work in Design (CSCWD)*, Nanchang, China, 2016, pp. 670–675, doi: 10.1109/CSCWD.2016.7566069.

18. G. Saravanan, Shailesh S. Parkhe, Chetan M. Thakar, Vaibhav V. Kulkarni, Hari Govind Mishra, G. Gulothungan, Implementation of IoT in production and manufacturing: An Industry 4.0 approach, Materials Today: Proceedings, vol. 51, Part 8, 2022, https://doi.org/10.1016/j.matpr.2021.11.604.

9 Implementation of Intelligent Manufacturing Using Cyber-Physical Systems

*Shreya Kagrawal, Sarvesh Tanwar,
and Shivangi Mishra*

9.1 INTRODUCTION

Production is considered the basis of a formed society. In Germany, Industry 4.0 is promising due to the convergence of information and communication technologies (ICT) and the introduction of the Internet of Things (IoT) and the Internet of Services (IoS). This phenomenon, recognized as the Fourth Industrial Revolution, is revolutionizing contemporary manufacturing facilities by converting them into intelligent entities. The Smart Manufacturing Leadership Coalition (SMLC 2011) in the US has embarked on proposals intended for the fabrication of an innovative manufacturing theory known as smart manufacturing. Under this paradigm, enhanced sensing, authority, modeling, and platform knowledge are widely accepted, and data-driven manufacturing information is applied in a complex way throughout the whole product manufacturing life cycle. Through the use of IoT in the manufacturing sector, the World Wide Web of Manufactured Things was researched in China. The encompassing ten-year national policy referred to as "Made in China 2025" has since recognized smart manufacturing as a crucial topic.

An organization's ability to be innovative, rational, and consistent is improved by the increasing unity of cyber components that follow physical ones. The advent of the Internet of Things (IoT) [1] and Internet of Services (IoS) in the manufacturing sector marked the beginning of Industry 4.0, also referred to as the Fourth Industrial Revolution, a process which will undoubtedly transform contemporary factories into intelligent ones. One of the unique characteristics of intelligent factories is their potential to improve manufacturing elasticity with the help of real-time configurable machinery. The CyberPhysical Systems (CPS) are considered a key technology for the Fourth Industrial Revolution, even if they employ a variety of technologies.

A CPS is a system for controlling physical entities that use computational elements. However, such amalgamation is not that new. CPS, or Cyber-Physical Systems, have been in existence for several decades and have evolved significantly over time. The

DOI: 10.1201/9781032630748-9

concept of integrating computational systems with physical entities dates back to the early days of automation and industrial control systems. As technology advanced, CPS applications expanded beyond industrial settings.

While CPS may seem like a recent development, it has a long history rooted in the integration of computational frameworks with physical entities. From early industrial control systems to modern IoT applications, CPS has revolutionized various industries [2], enabling more efficient, precise, and automated control over physical processes. The phrase "computer integrated manufacturing system (CIMS)" refers to a combined manufacturing system that combines computation and physical activities. CIMS has centralized control process which leads to some limitations such as lack of content realization, stretchability and seld-configuration within a factory setting.

In sharp contrast to CIMS, IoT and CPS tools like RFID (radio-frequency identification) and wireless sensor networks enable improved monitoring and administration of practical-creation processes on a never-before-seen scale.

9.1.1 An Outline of CPS

Around 2006, the National Science Foundation (NSF) in the US began using the term CPS, which was coined by Helen Gil [3]. The name "cyberspace" was devised by William Gibson in his groundbreaking book, *Neuromancer* [4]. CPS is a physically designed system that tightly incorporates "cyber" components (components with computing, communication, and control functions) with the physical environment to provide cyber-accessible services with a variety of options. These productions are now becoming remarkably driven in terms of reacting to more customers by utilising a set of the latest practices and knowledge. Some studies by scholars stated that enterprises are in need of resources that are tricky to duplicate in order to have an economical "flea market" edge. This can be achieved by making measured investments in cyber physical infrastructure [5] on the ground level, but this presents a challenge because it's likely that these cutting-edge technologies won't work well in guaranteed production scenarios.Additionally, industries are being forced to adopt cyber-physical production systems over conventional methods of production as a result of growing sensitivity to client needs and shifts in the market. Instead of focusing on union, CPS is actually fighting to achieve the convergence of the physical and the computer-generated.

- A sophisticated connection that enables real-time collection of data from the system's physical component and response into its cyber component.
- A well-planned data management with evaluation and processing control needed to create the cyberspace.

9.1.2 CPS-Related Terminologies

CPS is founded on long-standing procedures and prevalent demonstrations that are closely tied to its concept. Examples include embedded systems, wireless sensor networks (WSN), Internet of Things (IoT) [6], CyberPhysical Internet (CPI), Web of Things (WoT), and even Wisdom Networks (B2T). Stated differently, these ideas represent extra regular classes of CPS [7].

The administration of new approaches, in contrast to CPS, is mostly dependent on continuous dynamic reactions; hence, operational processes such as mode adaptation, error detection, rate restriction, and other concerns may go unchecked. It frequently occurs. Thus, it is impossible to monitor persistence, temporal recovery, and changes in parameters over time in a useful way.

WSNs are made up of many microsensor nodes that are placed in a specified region to inspect physical or conservational bounds. WSNs are an open-loop monitoring method with a static setup. Furthermore, the majority of WSNs suffer the issue of a restricted number of nodes. While CPS embraces not simply radars plus actuators and is recognized by locked-loop restriction, WSN expansion will aid in the approval of CPS [8].

The IoT was initially named in 1999. IoT prominently intersects with CPS. The most decisive change is that initially targets object recognition with CPS focuses on information exchange and feedback to control things. Currently, the realm of IoT comprehends the management and supervision of tangible entities. The Cyber-Physical Internet (CPI) idea was developed to tackle the current problem of interoperability among various CPS. To do this, a network that unites all CPS under one roof on a worldwide scale was constructed. The goal of the Web of Things (WoT), which was inspired by the Internet of Things (IoT) [9], is to seamlessly connect physical items to the Internet by utilizing web service capabilities that have been proved to work.

CPS embody the fundamental realm of opportunity and serve as a significant track of reasonable vantage within the innovation-driven economy of the twenty-first century. Consequently, CPS have garnered considerable notice from activity, the academic world, and governmental entities alike. Recognizing the immense potential of CPS to profoundly impact national interests, the National Science Foundation (NSF) has allocated substantial financial resources to facilitate transformative research and cultivate innovation in this domain.

Additionally, the European Commission has recommended that CPS studies be included in the Horizon 2020 program. According to studies from the President's Council of Advisors on Science and Technology (PCAST 2008), CPS is one of the six ground-breaking civil technologies propelling the US economy forward. It is regarded as a key area of opportunity and a means of giving the nation a competitive edge [10]. Various application domains have been the subject of specific research programs. In particular, the focus will be on smart manufacturing, which is enabled by CPS, and will be discussed further in the book.

9.1.3   CPS over Traditional Manufacturing Process

The influence of CPS and IoT on manufacturing plants is considerable. Presently, CPS and IoT concepts are widely employed across various industries, albeit their implementation in the manufacturing sector remains restricted. Research in other fields has mostly concentrated on the application of sensors that are incorporated in manufacturing machinery or goods that have RFID tags included in them. The assessment of records extracted from these devices is comparatively reduced. However, these systems can be effectively executed to investigate the Internet of Things (IoT) information [11], which can be developed in the supervisory-making process. Notably, cyber-physical machines continue to yield certain advantages over conventional production systems.

9.2 LITERATURE REVIEW

Scientists have been working at CPS for years to bridge the gap between computer systems and the real world. Various elements of CPS have been proposed since early 2008. CPS abandons the old and accepted system in which practices were shared, coordinated, and analyzed through interactions and workplaces. In recent years, researchers have shifted their focus to network productivity rather than other forms of productivity. Some authors have proposed CPS design training methods to achieve the goal of business information management and made many suggestions for CPS design for business [12], from technical level thinking to the connection of material to information management. The concept is also seen in the field of small flexible industrial manufacturing. Research has also developed control technology to contribute to cycle time in the planning of jobs in production (manufacturing arrangements) and distribution in industry.

9.2.1 CYBER-PHYSICAL PRODUCTION SYSTEMS

The German National Academy of Science and Engineering (ACATECH) has established seven centers on Cyber- Physical Systems (CPS) and 17 corresponding technologies. Key to these qualities is the ability to understand and analyze the physical environment using tools such as sensor fusion, pattern recognition, and situational awareness [13]. Additionally, the interaction between machines and humans, as well as the ability to extract knowledge from different backgrounds, will also contribute to the novelty of knowledge in machine learning.

9.2.2 CPS FUNCTIONING

The recognition of Cyber-Physical Systems (CPS) has been ongoing for the last few years. But in the physical and computational world there are many problems that need to be solved to achieve the best results. The International Institute of Automation has approved the ISA-95 design, which divides circuits into five levels. Level 0 refers to physical development, Level 1 uses physical development and management, Level 2 controls the monitoring and control of physical processes, and Level 3 develops products. Includes Level 4 work. Management of business activities. Some authors have proposed a CPS scheme called the 5C architecture, which also includes five layers. The process provides a step-by-step approach to developing and implementing a CPS, from data collection to final analysis and value creation [14]. The following sections provide a detailed description of each level as we approach CPS implementation.

9.2.2.1 Intelligent Connectivity

The original phase entails the exact and dependable collection of information commencing the machinery and its components. Managers, company systems like Enterprise Resource Planning (ERP) or Manufacturing Execution System (MES) systems, or sensors can be used to retrieve these facts. Sensor gathering is not the only important part of the process; data must also be transmitted to a mid-server so that a defined procedure may synchronize data formats and procurement methods.

9.2.2.2 Data-to-Information Transition

Data-to-Information Transition is a crucial process wherein significant information is derived from raw data. In contemporary times, a multitude of applications have emerged that possess the capability to forecast and verify the condition of various systems [15]. This development has primarily concentrated on the creation of algorithms specifically designed for this objective. Consequently, this level of conversion holds the potential to bestow a sense of "self-awareness" upon cyber-physical systems.

9.2.2.3 Cybernetics

Cybernetics plays a crucial role of data hub within the system of Cyber-Physical Systems. At this level, each machine's acquired information is stored, creating a network of interconnected machines. This network assists the extraction of additional information by comparing the personal condition of a machinery with the rack of the fleet, permitting for the scrutiny of execution and chronological data. This comparison facilitates the execution of predictive maintenance for the machines [16]. All phases are important for a CPS to operate well, but the cyber layer is a crucial link between lower levels' data collection and higher levels' decision-making processes. As a result, it is crucial that data analysis functions installed at this level operate effectively.

9.2.2.4 Cognition

Cognition describes the accomplishment of a complete understanding of the system under observation. To effectively transmit information to people, it is imperative that the most pertinent information is presented accurately. This level facilitates decision-making and allows for remote and cooperative diagnostics. Prioritising tasks in the maintenance process is made easier by the availability of comparative data and machine status information.

9.2.2.5 Configuration

The final stage of the configuration process consists of a feedback loop between the network domain and the system. Its main role is to maintain the machine and help it to repair itself. This level, which will help take corrective or preventive actions at the awareness level, which is the fourth level of the process, is generally called the Resilient Control System (RCS) [17].

5C architecture discussed the importance of vertical integration rather than horizontal integration. Based on this, this work presents an 8C architecture that incorporates three additional concepts (i.e., collaboration, customer, and context) into the existing 5C architecture [18]. The basis of the concept of integration is the incorporation of the value chain and the chain of different people involved in the production process. The concept of buyer focuses on the role the customer plays in the above process. Finally, context involves extracting, storing, and retrieving all product-related information, including but not limited to design, manufacturing, material product reviews, and after-sales service information.

9.3   CPS-BBASED SMART MANUFACTURING

The shift to more distributed and flexible production is driven by the need to be efficient, flexible and responsive to changing business needs. By breaking the traditional hierarchical structure, the company's operations can realize rapid changes, instantly adjust the production process, improve distribution and respond to customer needs [19]. One of the key benefits of this new approach is the ability to integrate and use information from a variety of sources, including sensors, machines, and other devices, to enable real-time monitoring, testing, testing and control of the manufacturing process. This allows companies to collect valuable information and make decisions based on it to increase productivity, efficiency, and overall performance. Circular autonomous and active units can also be used to increase the scalability and modularization of production systems. Each department can operate independently, make decisions based on local information, and coordinate with other departments when necessary, rather than relying on central control. This not only makes the entire process easier and more efficient, but also makes it easier for the system to expand and change production needs [20]. Additionally, using straight and loose joints reduces installation and repair costs by increasing material and design reuse. Companies can easily and quickly integrate new products into new production processes or existing products and technologies. This adaptability and flexibility also support the use of technologies such as artificial intelligence, machine learning, and robotics to increase the efficiency and effectiveness of production systems. Overall, moving from hierarchical management to a competitive, fair office environment brings many rewards, including greater agility, flexibility, scalability, and cost [21]. By using the principles of cyber-physical systems and offering autonomous and reusable units, companies will remain competitive in a dynamic and evolving business environment.

9.3.1   DISADVANTAGES OF CIM

Although CIM agreed to improve efficiency and competitiveness, it failed to deliver on its promises. The shortcomings of today's integrated production systems will reveal themselves in the next stage.

9.3.1.1   Dearth of Real-Time Data

This defect can be attributed to the absence of confirmation from Cyber-Physical Systems (CPS) [22], resulting in traditional manufacturing systems conducting in an open-loop manner. Subsequently, these systems have shown to be inadequate in terms of obtaining and administering real-time data, leading to a deficiency of critical manufacturing information that is either inaccessible or faltered. Accordingly, manufacturing enterprises find themselves lacking the necessary capacity to effectively adapt to uncertainties and modifications.

9.3.1.2   Information Island

Modern, complex products demand a myriad of production procedures, necessitating the vertical and horizontal integration of devices from shop floors to enterprise resource planning, embracing associated resources outside the boundaries of a single organization. Currently, existing solutions mostly use middleware and standards to overcome integration problems. Unfortunately, these solutions are usually insufficient since they lack the necessary flexibility and agility and cannot support a significant number of applications.

9.3.1.3   The Inadequacy of Intellectual Capacity and Proactive Behavior

Current control systems operate using a predetermined program and are limited to processing data that is preconfigured within a set cycle time. The allocation or management of functions typically occurs externally to the module through a fixed program, and their interfaces are rigid and unable to accommodate new or unfamiliar information.

9.3.2   Intelligent Manufacturing

The use of information and communication technology (ICT) [23] in business has transformed the production process through the application of smart manufacturing (SM) or cyber-physical production technology. This new approach attracted the attention of customers from business and academia as it promised to bring significant changes to the production process. The Office of Science and Technology Policy (OSTP) and the Office of Management and Budget (OMB) recognize the potential of smart manufacturing and support the need to support research and development (R&D). Their goal is to increase US leadership in key areas such as robotics, cyber-physical systems, and resilient manufacturing. By investing in this technology, the government hopes to boost economic growth and create more jobs. Smart Manufacturing Center (SMC) has many definitions that describe smart manufacturing as a technological revolution. These systems allow companies to quickly produce products, respond to changes in product demand, and optimize data and information. This intelligence and flexibility allow companies to operate more efficiently in a rapidly changing environment. The smart factory concept has become an important part of the Fourth Revolution, Industry 4.0.

Smart factory. This technology combines artificial intelligence, machine learning, and IoT technology to enable effective communication and collaboration between machines and humans. This combination provides efficiency, productivity, and ease of production. By integrating CPS into their operations, manufacturing companies can achieve high levels of automation, tracking, and time management. This not only increases the overall efficiency and effectiveness of the manufacturing process, but also leads to predictive maintenance and efficient resource allocation.

In summary, the use of ICT in business has led to smart manufacturing and this is expected to have a major impact on production. Advanced research and development (R&D) support from government agencies such as OSTP and OMB is critical to supporting US leadership in this field. Improving the application of advanced skills defined by SMLC enables companies to quickly respond to business needs and optimize their connected equipment products and services. Smart factory is an important part of the business.

9.3.2.1 Real-Time Access to Data

In today's industry, information access is essential as planning and scheduling procedures frequently lack comprehensive information. However, by utilizing several nodes, integrating digital computing systems and physical items smoothly, and sensing objects, Cyber-Physical Systems offer a solution to this problem. CPS gives decision makers the ability to acquire data in real time and access pertinent industrial information, enabling them to make more educated decisions. CPS also makes distributed intelligence for intelligent devices possible, allowing instant access to information at manufacturing sites. This makes it possible to gather input on the system or connected to production, which can help with closed-loop decision-making. Furthermore, uncertainties like performance indicators, traceability, and diagnostics can be rapidly incorporated into high-level system programming and control. As a result, manufacturing systems become more agile and robust.

9.3.2.2 Reconfigurable and Interoperable Abilities

The process of integrating heterogeneous systems (such as proprietary drivers or third-party connections) is associated with high costs and requirements. Additionally, point-to-point communication is inflexible. In contrast, cloud computing and network service technologies used in intelligent Cyber-Physical Systems are dynamic and reconfigurable. The organisations of these systems can be changed and used in autonomous control facilities thanks to Supervisory Control and Data Acquisition (SCADA) and Distributed Control Systems (DCS) capabilities. Through the development of service tilting method, modular design and open design, each device can easily express its function uniquely and integrate with other methods directly in the transition layer as required. This ability can be considered to constitute an intelligent CPS with a high degree of control. The creation of systems based on CPS can consist of subsystems where components are delivered as services that can be changed or removed. The operation of the system can be done by installing and configuring these services to complete the complex tasks that need to be done. The CPS-based reconfigurable manufacturing process can be easily adapted and seamlessly integrated into the new one. The integration and collaboration of these systems in production is not limited to low-level physical equipment but also includes advanced business management. Based on the advanced application of service orientation, the communication at the target level of the enterprise has achieved the relationship and change of business development. Adaptation can be done by transferring production to the web service provider's central service provider (e.g. to the cloud). All parts of the production body can be serviced by the system without changing the computer hardware.

9.3.2.3   Decentralized Managerial Making

In the business environment, there is a trend towards decentralization and communication, as opposed to traditional monolithic hierarchical or centralized structures. This change requires the introduction of flat, decentralized automation and next-generation service-centric automation. The next generation is paving the way for collaboration between gadgets and devices, regardless of physical location. Smart Machine (SM) ensures the service cooperation of the joint venture and make decisions autonomously. In addition, SM allows the business process to recognize and respond to events independently, unlike the business hierarchy of traditional applications. Use information, models, and analysis to predict events and monitor management to reduce the impact of risk and uncertainty. CPS manufacturing can use the Internet of Things to enhance standards-based measurement and defect detection to improve control of production processes [24]. This approach also enables efficient and flexible production.

9.3.2.4   Astuteness and Proactivity

Currently, there are several barriers to ICS adoption, including numerous device and system incompatibilities, difficult-to-access data, business contributions that are configured and utilized in a static manner, and reactive rather than proactive automation. Intelligent Manufacturing refers to the proactive process of combining a large number of distributed units into a collection of self-sufficient, fault-tolerant, reusable units through the use of intelligent control algorithms. This corporation is built on modular and scalable structures that allow it to act independently, take the lead in teamwork, and communicate with one another to achieve common goals.

Whereas machines acquire the capability to autonomously regulate their maintenance and repair approaches based on the level of workload they encounter. Additionally, they ensure the availability of backup capacities to sustain production in the event of interruptions caused by maintenance activities. Smart machines exhibit the ability to engage in the practice of "predict and prevent" rather than relying on a "fail and fix" approach. Numerous tools, including sensors, actuators, controls, and additional computer devices, are used to do this. Intelligent algorithms can also provide robots the capacity to self-regulate and adjust to complicated settings without the need for human assistance.

9.4   ARCHITECTURE

The architecture of the Intelligent Design System (IDS) , Cyber-Physical System (CPS) is divided into three layers: physical link layer, middle layer and layer. This section also explains the function and purpose of each layer [25].

9.4.1 PHYSICAL CONNECTION

Data collected by sensors can be used to improve the production process, increase product quality, and reduce time. For example, sensors can detect when the machine exceeds its normal operation and alert workers before failure. In addition, sensors can be used to monitor raw materials and finished products to ensure quality delivery to customers. Control sensors in the production process can also provide real-time tracking and monitoring of the production process. This allows rapid adaptation to change, such as changes in demand or changes in raw material quality. By using sensor data to improve production processes, companies can reduce waste, increase efficiency, and increase profits.

In summary, sensors portray a crucial role in the implementation of cCyber-Physical Systems in manufacturing. They have the caliber of enabling machines to perceive their environment, collect data, and communicate with other machines and systems. By utilizing data from sensors to optimize the manufacturing process, companies can advance product value, cut downtime, and strengthen profitability.

9.4.2 MIDDLE LAYER

It is suggested that the hyperphysical system's intervening stratum make it easier to extract and send data from the grounded components to the midway server for further investigation. The production instructions of the compute layers and the external applications—namely, quality control, dynamic work scheduling, and condition monitoring—help to act as control regulators. This layer serves as the connecting thread between the computing layer, physical connection layer, and external applications in the Cyber-Physical System. Several functions, which are described in more detail below, must be supported by the intermediate layer in order to implement the Cyber-Pphysical Ssystem.

   Device Administration: Several outside applications are utilized in Cyber-Physical Systems to manage devices from different brands, each adhering to their own unique standards and communication protocols. In order to achieve seamless integration and facilitate plug-and-play functionality within an industry, a public device module is required to effectively coordinate these diverse devices.

   Interface Definition: The data interface within the Cyber-Physical System (CPS) serves as a means to facilitate communication between nodes. Its purpose is to transmit data and information to external applications and the computational layer while concealing the intricacies of diversity.

   Data Management: This is important in many industries where information is retrieved from different products and RFID. Extracted data may include quality data such as tolerance, location, roughness, and dimensions, machine performance (such as vibration, speed and power), and the produced condition of the surface at specific times, noise, humidity, and temperature. To effectively manage and share this information in the internal market environment, an integrated information system that will provide different types of information and organizational structure is required.

9.4.3 LAYER

There is a lot of information, many types and models calculated by RFID and sensors from EIS (Enterprise Information System) such as SCM, ERP, and MES. Extracting this information requires a detailed process and representation that can provide a better understanding of the production process, the performance, and quality of the machine [26]. To explain this, let's consider the concept of planning and online evaluation on the shop floor, where the product code is integrated with data collected through data processing. This integration makes sense when machines are operating in a complex production environment. The calculator should solve the problem of flow calculation and batch calculation. Stream computing is used to distribute data streams from sensors, while batch computing focuses on processing large datasets based on historical data.

After streaming and bulk data analysis, the results are sent to the web machine for protection and limitation of operation. This process acts as a control, ensuring that the work environment is conscious and flexible. Use data mining techniques to leverage the shop environment to understand manufacturing processes and machine operations. These systems facilitate decision-making by integrating information obtained from human experience. The remainder of this chapter will discuss technologies that support Cyber-Physical Systems.

9.5   KEY ENABLING TECHNOLOGIES

Data-driven decision making is the reason for CPS's success in manufacturing. Technology involves the use of complex algorithms and machine learning to collect and analyze data from various sources in the production environment. By analyzing this data, CPS can improve the production process by making informed decisions and appropriate efficiency and productivity measures to increase efficiency. Another important technology to deliver CPS is the analysis of big business data related to the production process. As production technology becomes digital, more and more information is produced at every stage of the production process. Sometimes called "big data," this data contains patterns and insights that can be used to improve output. CPS can manage and improve the production process by using analytical tools and methods to analyze the relationship between different activities, analyzing the environment, identifying vulnerabilities, and predicting potential problems. To get desired outcomes in a CPS setting, a variety of machines, sensors, devices, and systems need to coordinate and interact with one another. To promote collaboration and information exchange amongst various instruments, connectivity refers to the establishment of effective and efficient communication routes. Collaboration, independent of product or technology, guarantees that various devices and systems function

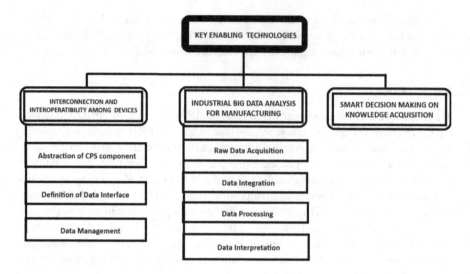

FIGURE 9.1   Key enabling technologies.

well together [27]. The technology unifies different components into a single CPS environment, facilitating efficient distribution, collaborative decision-making, and teamwork.

Figure 9.1 provides a visual representation of these three key technologies. It illustrates how intelligent decision-making, industrial big data analysis, and interconnection with interoperability are interconnected and work together to enable the successful implementation of CPS in manufacturing systems. The figure highlights the importance of these technologies in optimizing manufacturing processes, improving productivity, and achieving higher levels of automation and control. By considering and leveraging these fundamental technologies, manufacturers can unlock the full potential of CPS and drive innovation, competitiveness, and sustainability in their operations.

9.6   CONCLUSION

This chapter began by examining the similarities of Cyber-Physical Systems (CPS) to other situations and experiences, and also provided a brief overview of CPS. Provisions regarding the use of CPS designed to replace existing business processes were also considered. Opinions on producing CPS models with features such as rapid notification, revision, collaboration, decision-making, informing, and initiative are also discussed. This approach aims to overcome the recent limitations of structural design. The discussion also includes important support tools for machine design. It is conceivable that the integration of CPS into production can be considered an important factor in the advancement of manufacturing in the future [28, 29].

Researchers say CcyberPhysical Systems are quick solutions. Engineering research has now aroused great interest. Integrating CPS into system design is seen as a strong step for future engineering expansion. Whether it is the implementation of Industry 4.0 [30] or the integration of production with CPS, all of these are to support smart manufacturing and support the development and transformation of product development. These studies are designed to bridge the gap between different technologies, especially Cyber-Physical Systems and smart manufacturing. These conflicts arise from the lack of coordination and cooperation between these areas.

This research confirms the importance of developing technology to provide a general model and theoretical basis for the integration of smart manufacturing [31, 32]. It aims to provide a solid foundation for the use and advancement of production technology. However, it is important to understand that the purpose of this study is to develop a framework and a theoretical framework. Therefore, further research is needed to examine and understand the basic mechanisms of SCPS architecture. As a result, this chapter addresses the differences between IoT and IoS, CPS and smart manufacturing.This underlines the need for an architectural and theoretical framework for the integration of smart manufacturing. However, further research is needed to examine the characteristics of the SCPS architecture.

REFERENCES

1. Tanwar S, Paul T, Singh K, Joshi M, Rana A. Classification and imapct of cyber threats in India: a review. *In2020 8th International Conference on Reliability, Infocom Technologies and Optimization (Trends and Future Directions)(ICRITO),* ((2020): 129-135.

2. Atzori, Luigi, Antonio Iera, and Giacomo Morabito. "The internet of things: A survey." *Computer Networks* 54.15 (2010): 2787–2805.

3. Baheti, Radhakisan, and Helen Gill. "Cyber-physical systems." The impact of control technology 12(1), pp. 161–166, 2011.

4. Islam, M. S. (2021). Augmented Reality and Life in the Cyberspace in William Gibson's Neuromancer. *Advances in Language and Literary Studies, 12*(4), 30–34.

5. Bi, Z. M., and B. Kang. "Enhancement of adaptability of parallel kinematic machines with an adjustable platform." *Journal of Manufacturing Science and Engineering* 132.6 (2010): 061016.

6. Bi, Zhuming, Li Da Xu, and Chengen Wang. "Internet of things for enterprise systems of modern manufacturing." *IEEE Transactions on Industrial Informatics* 10.2 (2014): 1537–1546.

7. Marculescu, Radu, and Paul Bogdan. "Cyberphysical systems: workload modeling and design optimization." *IEEE Design & Test of Computers* 28.4 (2010): 78–87.

8. Davis, Jim, et al. "Smart manufacturing." *Annual Review of Chemical and Biomolecular Engineering* 6 (2015): 141–160.

9. De Amorim, et al. "Practical aspects of mobility in wireless self-organizing networks [Guest Editorial]." *IEEE Wireless Communications* 15.6 (2008): 6–7.

10. Derler, Patricia, Edward A. Lee, and Alberto Sangiovanni Vincentelli. "Modeling cyber–physical systems." *Proceedings of the IEEE* 100.1 (2011): 13–28.

11. Dillon, Tharam S., et al. "Web-of-things framework for cyber–physical systems." *Concurrency and Computation: Practice and Experience* 23.9 (2011): 905–923.

12. Dong, Jietao, Tianyuan Xiao, and Linxuan Zhang. "A prototype architecture for assembly-oriented cyber-physical systems." *AsiaSim 2012: Asia Simulation Conference 2012*, Shanghai, China, October 27–30, 2012. Proceedings, Part I. Springer Berlin Heidelberg, 2012.

13. Dubey, Rameshwar, et al. "The impact of big data on world-class sustainable manufacturing." *The International Journal of Advanced Manufacturing Technology* 84 (2016): 631–645.
14. Feld, T., M. Hoffmann, and R. Schmidt. "Industry 4.0—From smart products to smart production." *Information Management and Consulting* 27.3 (2012): 38–42.
15. Gunes, Volkan, et al. "A survey on concepts, applications, and challenges in cyber-physical systems." *KSII Transactions on Internet and Information Systems (TIIS)* 8.12 (2014): 4242–4268.
16. Hashem, Ibrahim Abaker Targio, et al. "The rise of "big data" on cloud computing: Review and open research issues." *Information systems* 47 (2015): 98–115.
17. Hermann, Mario, Tobias Pentek, and Boris Otto. "Design principles for industrie 4.0 scenarios." *2016 49th Hawaii international conference on system sciences (HICSS).* IEEE, 2016.
18. Huang, G. Q., P. K. Wright, and Stephen T. Newman. "Wireless manufacturing: a literature review, recent developments, and case studies." *International Journal of Computer Integrated Manufacturing* 22.7 (2009): 579–594.
19. Huang, George Q., Y. F. Zhang, and P. Y. Jiang. "RFID-based wireless manufacturing for real-time management of job shop WIP inventories." *The International Journal of Advanced Manufacturing Technology* 36 (2008): 752–764.
20. Yao, Xifan, et al. "Smart manufacturing based on cyber-physical systems and beyond." *Journal of Intelligent Manufacturing* 30 (2019): 2805–2817.
21. Koubâa, Anis, and Björn Andersson. "A vision of cyber-physical internet." *8th International Workshop on Real-Time Networks.* 2009.
22. Hsu, Cheng, ed. *Service enterprise integration: An enterprise engineering perspective.* Vol. 16. Springer Science & Business Media, 2007.
23. Lasi, Heiner, et al. "Industry 4.0." *Business & Information Systems Engineering* 6 (2014): 239–242.
24. Lee, Edward A. "Cyber physical systems: Design challenges." *2008 11th IEEE international symposium on object and component-oriented real-time distributed computing (ISORC).* IEEE, 2008.
25. Lee, Jay, Behrad Bagheri, and Hung-An Kao. "A cyber-physical systems architecture for industry 4.0-based manufacturing systems." *Manufacturing Letters* 3 (2015): 18–23.
26. Lee Ju Yeon, et al. "Ubiquitous product life cycle management (u-PLM): a real-time and integrated engineering environment using ubiquitous technology in product life cycle management (PLM)." *International Journal of Computer Integrated Manufacturing* 24.7 (2011): 627–649.
27. Lee, Jay, M. Ghaffari, and S. Elmeligy. "Self-maintenance and engineering immune systems: Towards smarter machines and manufacturing systems." *Annual Reviews in Control* 35.1 (2011): 111–122.
28. Lee, Jay, Hung-An Kao, and Shanhu Yang. "Service innovation and smart analytics for industry 4.0 and big data environment." *Procedia CIRP* 16 (2014): 3–8.
29. M. A. Ali, B. Balamurugan, R. K. Dhanaraj and V. Sharma, "IoT and Blockchain based Smart Agriculture Monitoring and Intelligence Security System," *2022 3rd International Conference on Computation, Automation and Knowledge Management (ICCAKM),* Dubai, United Arab Emirates, 2022, pp. 1–7, doi: 10.1109/ICCAKM54721.2022.9990243.
30. V. Sharma, N. Mishra, V. Kukreja, A. Alkhayyat and A. A. Elngar, "Framework for Evaluating Ethics in AI," *2023 International Conference on Innovative Data Communication Technologies and Application (ICIDCA),* Uttarakhand, India, 2023, pp. 307–312, doi: 10.1109/ICIDCA56705.2023.10099747.

31. A. Raj, V. Sharma and A. K. Shanu, "Comparative Analysis of Security and Privacy Technique for Federated Learning in IOT Based Devices," *2022 3rd International Conference on Computation, Automation and Knowledge Management (ICCAKM)*, Dubai, United Arab Emirates, 2022, pp. 1–5, doi: 10.1109/ICCAKM54721.2022.9990152.
32. Tanwar S, Kumar A. A proposed scheme for remedy of man-in-the-middle attack on certificate authority. *International Journal of Information Security and Privacy (IJISP)*,. 11(3), 2017 :1-4.

10 Use Cases of Intelligent Manufacturing

Shivam Goel, Alka Chaudhary,
and Vandana Sharma

10.1 INTELLIGENT MANUFACTURING

Fabricating constitutes the comprehensive preparation of collecting inputs, organizing, and changing them into the wanted yield. Shrewdly Fabricating speaks to a modern fabricating framework that synergizes the capabilities of people, machines, and forms to achieve the most favorable fabricating results [1].

The advancement of fabricating has been significant, moving past the early stages of Mechanical Insurgency. Present-day fabricating, not as it emphasizes quality and amount, but moreover prioritizes asset preservation and handling maintainability. Producers in the time of keen fabricating endeavor to keep pace with the continually extending showcase and the ever-heightening number of client requests through clever fabricating practices.

While ordinary fabricating depends on the existing information and mastery of mechanics, Advanced Manufacturing Systems empower dynamic learning from past generation information, understanding perplexing forms, anticipating generation results, and creating progressed alternatives.

Intelligent fabricating is regularly signified as 'smart manufacturing,' which encompasses fabricating forms that use mechanization, information gathering, counterfeit insights, machine learning, or other innovation techniques to set up ideal generation situations [2].

Manufacturers can progress past mechanizing assignments to executing end-to-end forms by amalgamating computerization advances like the Internet of Things (IoT) and Robotic Process Automation (RPA) with AI techniques such as computer vision, machine learning, and conversational AI. This amalgamation is commonly alluded to as shrewd computerization, cognitive mechanization, or hyper-automation.

10.1.1 THE IMPORTANCE OF INTELLIGENT MANUFACTURING

As the apex of accomplishments in the fabricating division rose amid the period known as the Mechanical Insurgency, human information and abilities confronted impediments in keeping pace with mechanical progressions. These winning impediments stemmed from administrators battling to remain side by side with the most recent innovative breakthroughs, rendering the ancient generation framework ill-equipped to handle instabilities and complexities in processes.

DOI: 10.1201/9781032630748-10

The ask for high-quality functions from a growing client base required raising era volumes and making strides to abdicate quality interiors in shorter timeframes while still managing era costs. The complement of resource optimization and environmentally considerate creation heightens weight on companies. Conventional methods proved inadequate in meeting objectives as human workers grappled with adapting to intricacies or memorizing vast amounts of data required for predicting output results [3].

Conventional manufacturing approaches additionally required operators to manually configure specific input parameters in machines and equipment before the commencement of production, relying on trial and error or the operator's experience. Despite years of expertise, machine operators might not comprehend every potential ambiguity or complexity that could arise during manufacturing.

Consequently, when a complication occurred, the entire process had to be halted and reconfigured to address the issue, incurring significant costs and inefficiencies in terms of time. These delays could have been mitigated if the system could record past uncertainties and recommend suitable input settings.

Hence, there emerged a need to integrate machines and new technologies to enable swift predictions and cultivate a flexible production method based on self-learning. The concept of an Intelligent Manufacturing system was conceived through the utilization of information technology.

10.1.2 AI AND THE FUTURE OF MANUFACTURING

In the manufacturing realm, AI holds the greatest potential for enhancing planning and operations on the production floor [4]. The ensuing are some notable applications of artificial intelligence in the manufacturing industry:

- Intelligent and self-optimizing machines automating manufacturing processes.
- Predicting efficiency losses to enhance planning.
- Identifying quality deficiencies to aid in predictive maintenance.

The subsequent sections will delve into the most noteworthy use cases within each of these categories. Referencing Figure 10.1, we can explore how AI is poised to revolutionize manufacturing.

10.2 PARADIGMS OF INTELLIGENT MANUFACTURING

The concept of Intelligent Manufacturing has undergone continual evolution over time to incorporate the latest technological advancements into the production system. Depending on the degree of information technology and its integration characteristics with manufacturing systems, Intelligent Manufacturing can be broadly categorized into three models or paradigms, as illustrated in Figure 10.2 [5].

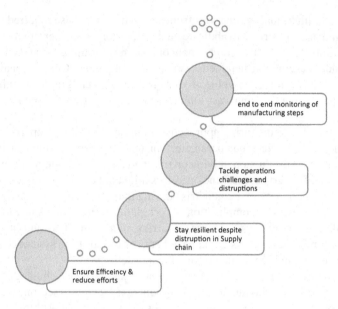

FIGURE 10.1 Transformation using AI.

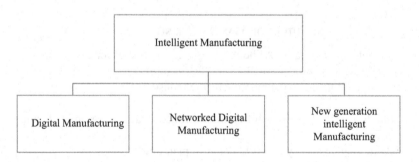

FIGURE 10.2 Paradigms of intelligent manufacturing.

10.2.1 DIGITAL MANUFACTURING

Also known as the inaugural phase of Intelligent Manufacturing, the digital manufacturing paradigm serves as the cornerstone upon which subsequent paradigms are built. In the 1980s, a pivotal shift occurred as production plans and output designs transitioned from traditional paper charts and graphs to computer-based platforms. Amid this stage, virtual reenactments of plant structures, item improvement, apparatus utilization, and workforce elements were produced to build up an optimized esteem chain. This approach pointed to diminishing costs, upgrading item quality, and maximizing asset utilization.

Through digitalization, the time required for making things custom-fitted to buyer needs has seen a noteworthy decrease. Companies presently use real-time information and reenactments to fine-tune generation forms, minimize squandering, and quickly adjust to advancing showcase requests. Computerized fabricating ~~not as it were~~ raises efficiency but also powers advancement by encouraging quick prototyping and customization of items. This worldview is an affectation, a transformation in conventional businesses, laying the foundation for a more feasible and active future in manufacturing.

10.2.2 Manufacturing on the Internet of Things

After reviewing the Netflix business model presented below, we will explore how it used the design thinking process, which includes the following stages:

1. Empathize: Unlike many cases where the research is not focused on Netflix's customers, this approach emphasizes the identification of various solutions by taking a well-rounded approach to investigation. Not only that, but the issue was also identified where customers faced difficulty when they did not rent from a physical movie store but had to go there in person. The subsequent increase in this issue was also highlighted.
2. Characterize: Immediately the thought of Netflix comes to mind as the team making an appear that portrays science arrangements, but the specific case is greater than fair that. Diverse from the over case the faceless shopping or virtual reality as well have as a result which is browsing at domestic or not requesting things you might never have considered ordering.
3. Ideate: Especially, there were a few thoughts proposed by Netflix included as the troubles seem be handled ideally. The (same as over) vetting/trial with membership benefit and the rental of DVD was moreover one of the key thoughts of shaping our benefit. The producer chose to send DVD players to the clients' homes specifically (without bundling them with any other gadget). Here, this specific facilitating gave him another opportunity of sparing for videotapes, essentially to the degree that he was not obliged to lease the firmly estimated tapes from standard rental motion picture stores.
4. Model: At last, Netflix centered on the most advanced model useful for its as of now existing stage which is at risk and energized future changes and quick modifying. They might have selected for the DVD benefit as the to begin with development, opening a couple of them and expanding their number of clients; at that point, once the last mentioned begin accepting the benefit, duplicate the number of those benefit units.
5. Test: Netflix must counsel its administrative specialist or maybe than prescribing alone together with its fundamental control on its compliance and nonstop evaluations of the accomplished comes about. What this implies, basically, is that tips for upgrading the circumstance of clients ought to stay the most vital figure for the company to proceed. Hence, the organization chosen differentiating to being eradicated from the advertise by the current patterns, it may take after a distinctive approach. The firm has said what kind of commerce would be comparable in the future that citizens will inquire for a arrangement of moment administrations if any and would set up the company considering the viewpoint.

Netflix's item group wrapped up culminate creative sheet where adjust of the different sorts of classes and motion pictures was associated with their watchers piece of intellect. Firstly, we created the touchpage to meet the necessities and interface of this gathering of people. Furthermore, all the substance of our occasions (subjects and motivation) were created to be engaging to the same target showcase. Fundamentally, since the initiation of Netflix one year prior, we can't offer assistance but address whether Hulu would still be as fruitful if it were established at that point. In 2011, Hulu made its attack into the Original TV arrangement. Be that as it may, by December of 2016, it changed its stage into one oversimplified client interface.

Their improvement is ongoing based on the cycle plan considering forms that ordinarily include all sides and take the showcase target's audience into thought. These inputs are at that point viably utilized in repairing the beginning issue and come up with models and arrangements. By means of the methodology which was driven by clients, the enterprise took leads the showcase and proceeds to see for different sources of specialties. The last mentioned is conceivable much appreciated to developing a faithful customer-base who seeing a modern request in offers.

10.2.3 INTELLIGENT MANUFACTURING OF THE FUTURE

The integration of manufactured insights, artificial intelligence (AI) with advanced and organizing advances, has yielded critical strides in Cleverly Fabricating Frameworks (IMS). The application of AI in mechanical forms holds the guarantee of a significant update of generation strategies, possibly rendering human judgment skills obsolete. Next-generation Intelligent Manufacturing, empowered by AI, can autonomously undertake research and development (R&D), devise novel processes, designs, products, and even establish innovative business models, all without human intervention. Leveraging the efficiency of robots over human counterparts, AI not only reduces the manufacturing time but also streamlines the innovation and ideation processes.

The models are not isolated from each other but rather they leverage and expand upon the strengths of the preceding model.

10.2.3.1 Evolution in Manufacturing

Intelligent Manufacturing is intricately tied to the concept of the Fourth Industrial Revolution, commonly known as Industry 4.0. Crucial components of Industry 4.0, including artificial intelligence, machine learning, the Internet of Things (IoT), and cloud computing, play integral roles in enabling Intelligent Manufacturing. Fundamentally, Intelligent Manufacturing is characterized by the real-time communication among interconnected devices utilizing embedded sensors and cloud-based software. This, in conjunction with machine learning and robust data analytics, establishes a heightened level of visibility and flexibility [11]. In essence, Intelligent Manufacturing represents the synergy of various components working harmoniously to create a more comprehensive and informed whole. When employed collectively, these techniques empower manufacturers to adopt a proactive stance, anticipating challenges rather than merely reacting, and allocating more time to forward-thinking strategies.

Intelligent Manufacturing (IM) specifically involves leveraging the combined intelligence of individuals, processes, and equipment to influence production economics. The overarching objective is to optimize manufacturing resources, enhance business value and safety, and eradicate waste, both on the shop floor and in back-office operations. This is achieved through the utilization of cutting-edge tools such as the latest manufacturing execution systems (MES), intelligent devices, machine-to-machine connectivity, and data analysis across production lines and facilities, all while ensuring the fulfillment of customer delivery and quality objectives.

10.3 THE INDUSTRIAL REVOLUTION'S EVOLUTION (FROM INDUSTRY 1.0 TO 5.0)

The inaugural stage of the Mechanical Transformation commenced in the eighteenth century and experienced five progressive stages as innovation and forms proceeded to development in the consequent decades

10.3.1 INDUSTRY 1.0

The onset of the Mechanical Transformation, commonly alluded to as Industry 1.0, checked an essential milestone in human history. Started in the late eighteenth century and holding on into the early nineteenth century, this transformative period, on a very basic level, changed the strategies by which society made merchandise. Characterized by the far-reaching appropriation of mechanized fabricating forms, steam control, and the foundation of the earliest-known manufacturing plants, it implied a move from agrarian and craft-based economies to industrialized ones. Critical developments like the spinning jenny and the steam motor played instrumental parts in empowering mass generation and increased proficiency. Industry 1.0 activated critical financial shifts, counting urbanization, as people moved from rustic settings to urban centers in interest of business in production lines. This period laid the basis for ensuing mechanical insurgencies and laid the foundation for the modern mechanical scene, underscoring the significant part of innovation in forming the direction of human advancement.

10.3.2 INDUSTRY 2.0

Industry 2.0, commonly known as the Moment Mechanical Insurgency, spoke to an essential point in the advancement of fabricating. Unfurling in the late nineteenth century and amplifying into the early twentieth century, this period saw the broad selection of power, the beginning of the assembly line, and the utilization of features. It stamped a significant flight from manual labor and steam-driven apparatus to more robotized and zapped frameworks, essentially improving the proficiency of fabricating forms. The appearance of power revolutionized manufacturing plant operations, empowering steady and effective control for machines.

The assembly line, attributed to Henry Ford's automobile-production developments, revolutionized mass generation by breaking down complex errands into less difficult, specialized occupations performed by either people or robots. The consolidation of conversely parts assist streamlined generation, disentangling the substitution and repair of components.

Beyond raising efficiency, Industry 2.0 affects significant social and financial changes. It fueled the development of urban centers, created more job openings, and contributed to the development of the center course. This era laid the foundation for the modern fabricating scene, setting the scene for ensuing mechanical insurgencies and the innovative headways that characterized the twentieth century.

10.3.3 Industry 3.0

Industry 3.0 recognized as the Third Mechanical Transformation, spoke to an essential role in the direction of fabricating and computerization. Arriving in the late twentieth century, it was characterized by the broad joining of computer innovation and computerized headways into mechanical operations. A characterizing advancement of Industry 3.0 was the approach of computer numerical control (CNC) machines, presenting increased accuracy and mechanization to fabricating processes.

These machines seemed to be fastidiously modified to execute perplexing errands with extraordinary exactness, lessening the dependence on human labor over different businesses. The advancement of mechanical autonomy and the utilization of computer-aided design (CAD) assist streamlined generation forms. Industry 3.0 laid the groundwork for upgraded productivity and efficiency in fabricating, building up the basis for the consequent Industry 4.0 insurgency. In this following stage, the integration of the web, enormous information, and fake insights would usher in more modern and interconnected frameworks inside the mechanical landscape.

10.3.4 Industry 4.0

Industry 4.0 or the Fourth Mechanical Transformation, is stamped by significant changes in the fabricating and generation scene. This time is recognized by the integration of sophisticated advances and data-driven arrangements over different businesses, proclaiming phenomenal levels of computerization, network, and insights. Inside the domain of Industry 4.0, shrewd industrial facilities used innovations like the Internet of Things (IoT), fake insights (particularly computer vision), enormous information, and cloud computing to build up profoundly productive, flexible, and versatile generation systems.

These interconnected frameworks empower real-time observing, prescient upkeep, and data-driven decision-making, contributing to improved proficiency, minimized downtime, and increased item quality. Industry 4.0 not only expanded efficiency but moreover fortified development by encouraging customization and fast prototyping, on a very basic level, changing the approaches to item plan and fabricating. This mechanical insurgency amplifies its impact past fabricating, branching into segments such as coordination, healthcare, and agribusiness, rising as a principal driver of financial and innovative alterations in the twenty-first century.

10.3.5 INDUSTRY 5.0

As we usher in the period of Industry 5.0, an outstanding worldview shift in fabricating is seen, putting a noteworthy accentuation on the consistent integration of human aptitudes and imagination. Whereas Industry 4.0 prioritizes robotization and data-driven forms, Industry 5.0 moves to reintegrate people into the fabricating circle, leveraging their particular capacities to address perplexing issues and make nuanced choices. This rising mechanical transformation recognizes the irreplaceable part of human instinct, enthusiastic insights, and inventiveness in the fabricating domain.

The collaboration between people and smart machines, encouraged by advances like counterfeit insights, increased reality, and collaborative mechanical autonomy, encourages more adaptable and versatile generation forms. Industry 5.0 advocates for a more comprehensive and different workforce, cultivating advancement, customization, and proficiency in fabricating, all while maintaining a solid commitment to maintainability and moral hones. In this time, producers are not exclusively computerizing for effectiveness; instead, they are grasping the total range of human potential to impel the industry toward a more adjusted and feasible future, a concept assist investigated in the ensuing discussion.

10.4 HOW INTELLIGENT MANUFACTURING DELIVERS THE GOODS

Various and differing focal points can be realized through Brilliantly Fabricating [7]:

- Accelerated item improvement: The capacity for self-improvement expands past people, as shrewd and interconnected things, along with their advanced partners, give fundamental information and bits of knowledge to improve future versions.
- Enhanced visibility and connectivity: These are crucial elements for building a more resilient and disruption-resistant supply chain.
- Reduced working costs: Brilliantly fabricating analytics and real-time information enhancements enable streamlined operations, minimizing squandered time and exertion, and optimizing asset arrangement, driving to significant realized savings.
- Heightened security: Progress in mechanical autonomy innovation empower the appointment of hazardous assignments, once taken care of specifically by people, to independent or remotely worked gadgets. The capacity to execute errands in a computerized or virtual environment contributes to guaranteeing the security of cutting-edge laborers, particularly in times of social isolation.
- Improved resource execution: The integration of cleverly fabricating units with prescient support upgrades generation by anticipating and avoiding issues, avoiding delays caused by startling gear breakdowns.

- Empowered workforce: Cleverly fabricating computer program optimizes operations, advertising staff expanded perceivability to boost efficiency. Robotization of essential strategies permits representatives to center on higher-level assignments, and the application of shrewd network addresses the expertise by giving collaboration apparatuses and farther master help to direct laborers through investigating and best practices.
- Enhanced productivity: Automation of basic procedures allows employees to concentrate on more advanced tasks. In response to the evolving nature of manufacturing, where a skills gap has emerged, smart connectivity can bridge this gap by offering technicians and workers collaboration tools and remote expert assistance, guiding them through troubleshooting procedures and best practices.
- Improved customer satisfaction: Enhanced visibility and control over the production process help prevent issues leading to client delivery delays. Microsoft's best placement among unique companies is combined with well-managed resources to ensure products meet customer needs.

10.5   USES OF INTELLIGENT MANUFACTURING

10.5.1   SUPPLY-CHAIN MANAGEMENT

By identifying key elements for success, significance, and diversity, researchers demonstrate the important role humans play in changing the product. By addressing innovations such as the Internet of Things (IoT), artificial intelligence, and analytics, we can use real-time data and public information across the business to improve decision-making.

An important area where progress is being made in distinguishing between materials is decision-making. Here our ability to analyze data transactions, failures, backup plans, and external changes differs from our future needs. This capability is important for improving product quality, reducing shipping costs, and avoiding inventory problems such as low or overstock. Additionally, a great strategy allows us to adjust and change the plan as needed to adapt to the changing situation. The purpose of this change is not only to speed up shipping but also to reduce possible delays and ensure timely delivery of the product.

Quality control is another important part of the supply chain and smart manufacturing is important here. Using fraud detection and machine learning, we can identify patterns over time, distinguish reliable data from inconsistencies, and predict potential problems that have worsened in the past. This ensures that the product is presented properly and remains sharp throughout the shipping process. This diminishes misuse but also moves forward client satisfaction.

In the space of coordination, the integration of IoT and GPS watching with AI calculations enables real-time encounters in the stream of commodities over the supply chain. This empowers moved forward course optimization, expanded security, and more practical organization of transportation costs. AI advancement, reinforced by IoT sensors, is continuously utilized in organizing comprehensive supply chain shapes and counting assignments from stock organization to texture stacking and movement. This integration of AI, machine learning, computer vision, and mechanical innovation handles mechanization, and helps and refines supply chain organization shapes, making them more capable and cleverer.

10.5.1.1   Creates a Robust Communication Connection across Departments

Cutting-edge supply chain organization courses of action for the mechanical division reliably combine diverse divisions of a company into a bound-together arrange. Hence, this builds up a perfect communication channel among bunches, contributing to the change of by and expansive commerce performance.

10.5.1.2   Logistics and Warehouse Management

Within the dissemination center organization, sensors enabled by the Internet of Things (IoT) can successfully screen stock levels and conditions in real time. These sensors furnish correct data on stock levels, and temperature, and undoubtedly rack life, engaging control and lucky energizing. Calculations fueled by AI can intentionally optimize the course of activity of stock, minimizing travel time and reducing botches in the picking and squeezing of shapes. Moreover, computerization advancements like Robotized Guided Vehicles and meanders play a portion in texture taking care of, progress streamlining operations in interior stockrooms. As a shared space, productivity greatly increases course planning and tracking calculations, considering factors such as action conditions, climate, and movement windows, can analyze and optimize movement courses. The amalgamation of real-time GPS taking after and IoT sensors on cargo ensures reliable checking of the zone and condition of things, hence moving forward the security of shipments and empowering action responses to any issues that may develop in the midst of travel.

10.5.1.3   Development of Autonomous Vehicles for Logistics

The introduction of autonomous vehicles for coordination, energized by adroitly creating, stands as a groundbreaking progression in supply chain organization. These driverless vehicles are prepared with progressed sensors, counterfeit insights, and real-time data to guarantee steady and precise item conveyance. Shrewd fabricating plays a critical part in this change, empowering the integration of independent vehicles into the supply chain.

These instruments can end up as a fundamental portion of a more extensive organization where information investigation and future-proofing make strides, courses and discuss travel arranging, decreasing travel times and costs. Also, fabricating greatness bolsters the upkeep of these autonomous devices, predicts glitches and guarantees nonstop collaboration. The concept is complex: It consequently increments shipping speed, decreases the probability of blunders, diminishes fuel utilization and decreases carbon outflows. It moreover increments the perceivability of the supply chain, permitting merchants and exporters to track their shipments.

Shared instruments, guided by shrewd generation, have the potential to revolutionize commerce by making it more energetic, proficient, and conventional. As this development advances, the conveyance of items will be reexamined, eventually profiting both businesses and consumers.

10.5.1.4 Forecast Product Demand

Accurately evaluating necessities is a critical portion of successful supply chain administration and setting up a commitment system is a critical portion of moving forward with precision and execution. This handle employments progressed innovations such as human recognition and information examination to analyze real information changes, find designs, etc. It supplies the needed information. This not only implies that the company can make strides at the item level, but also guarantees that generation and conveyance can be balanced and meets the real needs of customers.

Smart fabricating goes past fundamental inquire about to coordinated variables such as progression, geological differences, and outside components such as climate or back. This best approach permits organizations to change their items and plans. In this manner, by encouraging straightforward and exact conveyance, businesses can diminish transportation costs, decrease the hazard related with out-of-stock or high-volume items, and increase client fulfillment. In this way, shrewd fabricating gets to be a vital device for item masters who endeavor to calculate and meet item needs effectively, delightfully, and efficiently.

10.5.1.5 Price forecasts

A great understanding of the introduction bolsters assist learning and critical determining decisions.

Adaptive determining methods: Leveraging real-time data, companies can utilize proactive estimating strategies that permit them to swiftly alter costs, competition, or other markers when there are request changes. This approach maximizes effectiveness while keeping up competitiveness.

Effective taken a toll administration: Through progressed investigation, you can recognize potential taken a toll investment funds in materials. This incorporates optimizing obtaining information, conveyance, and stock administration; this leads to lower costs and more unsurprising competition.

Customer-focused determining: Viably controls the organization's capacity to tailor estimates to person clients or particular item offerings. This alter increments client fulfillment and builds trust.

10.5.1.6 Robotics in Manufacturing

The combination of innovation and productivity in generation and supply chain administration is a combination that is changing the nature of present-day commerce. Mechanical independence empowers computerization and exactness in the generation prepare, in this manner expanding the capacity and steadiness of item quality. When these robotization systems include fabricating perceivability and IoT sensors, they can send information in real-time to give supply chain administration. The combination of shrewd fabricating and innovation is producing incredible intrigue in the world of supply chain administration. To begin with, it energizes changing generation arranging based on unused request data to guarantee items are delivered when they are required. The flexibility of the robot, which can be rapidly reconstructed or adjusted to oblige changes to make prerequisites, makes a difference in decreased lead time and makes everything easier.

10.5.1.7 Customer Management

Smart fabricating is changing the way companies track items and give clients with data almost their items. By coordination, modern advances such as fake insights, IoT, and information analytics, companies can make strides in client administration at distinctive portions of the supply chain. A vital space is request estimating. Companies can react to client needs utilizing counterfeit insights calculation and information examination. This allows them to alter their era and stock organization completely with legitimate client needs, in this way reducing lead times and ensuring thing openness when clients require them. The made strides precision in ask assessing not only lifts client fulfillment but additionally optimizes resource task and diminishes costs related with expect stock or last-minute era. Cleverly creating is driving positive changes in client relations all through the supply chain. Cleverly creating contributes to the overhaul of thing taking after and traceability interior the supply chain. Through the integration of IoT sensors and blockchain advancement, real-time perceivability into the status and zone of things can be finished, giving clients a straightforward manner the capability to screen their orders. This not only develops accept but as well empowers the proactive assurance of any potential issues in the midst of travel, driving to a more reliable and dependable client experience.

10.5.1.8 Inventory Management

Brilliantly creating plays a pressing part in changing stock organization interior of the framework of supply chain organization. Organizations can make more proficient and compelling item administration by utilizing progressed advances such as the Internet of Things (IoT), fake insights (AI) and information analytics. Perfect item taking care of is fundamental for the steady operation of the generation process.

Leveraging machine learning models and stock information, savvy mechanization can offer assistance producers the ability to dodge item blackouts and diminish waste:

Fast laborers when stock is low
Gauge required inventory
Illuminate representatives around new inventory

Inventory control mechanization also has other preferences. With an AI-powered framework, organizations can decrease the require for manual intercession by altering the cycle prepare based on predefined criteria. This basic handle not only decreases the possibility of human mistake, but moreover permits human assets to do a much better job.

10.5.1.9 Forecast product demand

Predicting the require for the right items is vital for compelling supply chain administration, and shrewd fabricating frameworks play a critical part in creating this capability. Utilizing advances such as fake insights and information analytics, these frameworks were best for giving bits of knowledge by analyzing verifiable deals information, showcase patterns, and different factors. This permits companies to create high-quality items and guarantee that generation and dissemination are based on customers' genuine needs.

Smart fabricating goes past examination and takes into account regularity, territorial varieties, and other variables such as climate or industry. This approach permits organizations to streamline their supply chains and generation forms, decrease carrying costs, and diminish dangers related with out-of-stock or high-volume items. Eventually, this increases client fulfillment through convenient and exact support.

Smart fabricating is a vital device for commerce experts that permits them to foresee and meet item needs proficiently and effectively [6].

10.5.1.10 AI for Order Management

Order administration frameworks must illustrate speed, proficiency, and adaptability to changes in advertise conditions, necessities, client needs or plan techniques. Designers may select to utilize AI-based frameworks or bots for the taking after purposes:

Optimizing the arrange section prepare and robotizing dreary tasks.
Powerful apparatuses to look for items and naturally make buy orders.
Solve complex issues related with diverse arrange sorts over different pipelines.
Improve the productivity and straightforwardness of stock arranging and arrange management.

The organization diminishes passage expenses and ensures benefits. AI-powered chatbots and virtual associates are getting to be progressively prevalent for client back and questions. These apparatuses give moment input that makes a difference increment client engagement and progresses the general requesting experience.

10.5.2 Predictive Maintenance

Predictive support driven by keen fabricating speaks to a transformation in supply chain administration. Utilizing progressed advances such as the Internet of Things (IoT), artificial intelligence (AI), and a great amount of information analytics, companies can move from checking applications to great experiences and forecasts. This adaptability permits them to foresee gear or resource disappointment and arrange support or substitution at the right time, decreasing downtime and lessening equipment-related costs. For supply chain administration, this implies credibility and resilience.

Integrating keen sensors into machines can screen the status and execution of basic hardware. Counterfeit insight calculations that right away handle this information can identify designs and imperfections that may demonstrate potential issues. These highlights permit for opportune intercession, decreasing the probability of unforeseen events and their affect all through the supply chain. Prescient support not only diminishes the effect of work, but also disentangles the allotment of resources.

By centering on support, organizations can expand the life of resources, decrease support costs and optimize resource allotment, eventually resulting in a great result. In spite of the fact that the innovation supporting virtual twins is progressed, the concept is direct: make a virtual reproduction of an item, resource, or property that serves as its computerized partner to improve the unique through optimization. NASA spearheaded this approach by refining shuttle demonstrated reenactments. Once an advanced twin is made, a ceaseless stream of information sets up an association between the physical and computerized partners: experiences assembled are connected to persistently make strides with the physical partner, whereas information from the real-world twin is utilized to refine the computerized partner ceaselessly. This comes about in a progressing cycle of communication and optimization that helps in conveying prevalent items, streamlining operations and costs, and making remarkable client experiences.

For illustration, an airline company might coordinate different information streams to create a computerized representation of an airplane. Also, the airline may build up an energetic information biological system to advance a responsive computerized demonstration of any flying machine, analyzing its wellbeing, effectiveness, and comprehensive history, involving all person components. This approach helps their clients in improving fuel productivity, decreasing support costs, and lifting general airline fleet readiness.

10.5.3 Cobots Working with Humans

Collaborative robots, known as "Cobots," talk to a transformative perspective of brilliantly creating, working with adjacent individuals to promote effectiveness, security, and capability over contrasting mechanical circumstances. These present-day robotized systems are especially made to offer assistance points of their callings. In the manufacturing section, cobots find growing applications in assignments such as get-together, quality audit, and texture managing. Their ease of programming and recreating works along with their adaptability, allowing them to rapidly alter to progressing era requirements.

10.5.4 RPA for Printed Material Automation

Factories commonly deal with a critical entirety of printed fabric, wrapping orders, requesting, and reports. Executing these assignments physically is time-consuming, and tions. The joining of Mechanized Plan Computerization into brilliantly creating suggests a basic progress toward fulfilling steady and beneficial operations interior in a present-day mechanical environment.

When associated to printed fabric computerization in the setting of brilliantly manufacturing, RPA extends the integration of computerized and physical shapes, optimizing diverse highlights of era and the supply chain. RPA can energize the digitization and computerization of paperwork-related assignments in manufacturing, covering works out such as data segment, report sorting, workflow organization, and quality control. This not only diminishes the likelihood of human error but moreover streamlines operations, ensuring expanded effectiveness and cost-effectiveness. The collaboration between RPA and brilliantly fabricating not only optimizes inner forms but moreover reinforces the capacity to react to showcase changes and client requests. It locks in makers to be nimbler and more flexible, contributing to more viable and competitive operations. Consequently, the integration of RPA for printed fabric computerization with smart manufacturing lays the establishment for a more savvy, interconnected, and profitable mechanical landscape.

10.5.5 Lights-Out Manufacturing Plants Spare Money

Lights-out generation lines, recognized by their irrelevant or add up to nonappearance of human intervention, epitomize the embodiment of mechanization interior the space of cleverly creating. These state-of-the-art workplaces saddle cutting-edge developments, tallying fake bits of knowledge, mechanical independence, and the Internet of Things, enabling them to work ceaselessly.

Primarily, it entirely diminishes labor costs as there is no require for on-site human pros; and operations can proceed ceaselessly. Blocked off upkeep and supervision progress contribute to gotten diminish. Moreover, warm control plants increment generation. The machine works without breaks, changes or exits, always produces vitality and indeed produces power. This increments utilization and decreases downtime, eventually advancing in general get to to capital and expanding benefits.

10.5.6 Visual Inspection

Integration of visual inspection and smart manufacturing is an essential part of today's equipment that uses technology to increase control and efficiency. Visual analysis has made significant progress in the Industry 4.0 era, which is the basis of decision-making with computers and information.

Currently the main marketing process continues to combine computer vision, machine learning, and the creation of emotions to transform the message consistently and quickly from advertising. These systems use high-resolution cameras, sensors, and image processing to identify any component or components in the product. This new approach is particularly useful for practical applications as it indicates rare omissions or deviations that would normally escape human observation.

10.6 DESIGN ITERATIVELY

Mechanistic learning algorithms are important in the design processes of engineers. For instance, designers or engineers input a set of certain parameters such as materials, sizes, weights, endurance loads, and manufacturing limits into software designs. Hereafter, the software produces several models for multiple designs on the same product. This allows more options for designing and enhances the design process through technology. The term Industrial Internet of Things (IIoT) is used to refer to this concept in business. You can enhance accuracy and productivity by integrating IIoT into your business.

10.6.1 INCREASED PRODUCTION SECURITY

Smart manufacturing requires integration of data and operations, and virtualized IT/OT architecture can ensure the security and efficiency of the manufacturing process. This integration not only improves the stability of production but also supports the stability of the entire industry. Unlike simple information silos, an integrated production environment supports visibility of information security and protection of intellectual property rights. In the context of improving production safety, smart manufacturing has the following advantages:

1. Integration of information and operations: By integrating information and operations into a unified system, smart manufacturing can reduce the risk of illegal security and ensure the integrity of the production process.
2. Continuous monitoring and analysis: Smart manufacturing systems can analyze the system and analyze production data to detect suspicious or potential security threats at an early stage.
3. Secure data sharing: Smart production supports data sharing with stakeholders through secure communication, while also protecting the confidentiality and integrity of the people involved.
4. Strong network security measures: Smart manufacturing protects valuable assets and sensitive data with strong network security measures such as encryption, authentication, and management.
5. Adaptive security protocols: Smart manufacturing systems adopt adaptive security protocols that can dynamically respond to changing network threats and vulnerabilities.

More importantly, the use of smart equipment not only increases productivity and efficiency, but also improves stability and ensures safety. Productivity thrives and evolves in a digital and continuous environment.

10.6.2 AI FOR PURCHASING PRICE VARIATION

Changes in raw material prices have an impact on profitability in businesses. Accurately estimating these costs and identifying the right supplier can be difficult. To effectively solve this problem, companies can leverage solutions that use proprietary technologies for monitoring and analysis.

There is a way to solve the price variation of technology-focused dashboards:

Features such as volume, diameter, material, and coating embody the same characteristics as design fear.

Manufacturers can use products such as country of origin, brand name, or functional indicators to prepare suitable products.

The tool can also use historical data and business models to estimate purchase prices and provide insight into changing costs.

Additionally, technology can help with price competition by creating benchmarks to compare prices from different sellers.

In addition to price estimation and price analysis, the solution also facilitates tracking of products from multiple suppliers.

Customer management: This centralized approach makes it easier to manage information, improves overall results and increases the efficiency of various production processes. This mechanism improves the storage and increases the productivity of the system.

10.6.3 AI Product Development

Integrating AI-driven product development can help companies conduct more testing and testing through augmented reality (AR) and virtual reality (VR) before starting the manufacturing process. Therefore, manufacturers will be able to reduce costs associated with trial and error.

- Accelerate the speed at which products reach the market.
- Enable engineers to anticipate and preempt potential issues before product launch.
- Streamline maintenance and debugging procedures.

Manufacturers can elevate and expedite their innovation through AI-centric product development, leading to the creation of novel and more advanced products that outpace competitors. Additionally, the continuous feedback loop inherent in AI algorithms enhances their performance with each iteration, thereby contributing to the evolution of superior goods.

10.7 REAL-WORLD EXAMPLES OF AI IN MANUFACTURING

As per the analysis by MarketsandMarkets, the worldwide market for artificial intelligence in manufacturing is projected to reach a valuation of $16.3 billion by the year 2027, exhibiting a robust compound annual growth rate (CAGR) of 47.9% from 2022 to 2027. Below are tangible examples illustrating the practical application of AI in the manufacturing sector [8,9].

1. **GE employs AI to shorten product design cycles.**

Engineers at General Electric leveraged AI technology to design tools that can expedite the manufacturing of aircraft engines and power turbines. Previously, GE specialists dedicated nearly two days to analyze fluid dynamics within a single tur-bine blade or engine component design. However, through the implementation of machine learning, experts at General Electric's research center in New York devel-oped a model capable of assessing a million design modifications in just 15 minutes. This advancement significantly accelerates the company's progress in creating its next generation of products [5].

2. Toyota works with Invisible AI.

Toyota collaborated with Invisible AI to incorporate computer vision technology into its North American facilities, leading to enhanced safety, quality, and efficiency in its production processes. Through the analysis of individuals' walking patterns, AI technology can predict and prevent minor faults and injuries. Notably, these AI solutions possess the capability to autonomously learn. Additionally, these devices operate independently without the need for a cloud or internet connection, featuring advanced technology and cameras to monitor activities on the factory floor [6].

3. PepsiCo streamlines production quality control.

Suntory PepsiCo encountered challenges in their soda plants where labels, includ-ing manufacturing and expiration dates, needed accurate reading. Due to occasional smearing caused by applying tags before the surface dried, production delays and costly shutdowns ensued. To overcome this, Suntory PepsiCo sought the expertise of Pacific Hi-Tech, which introduced a "Machine Vision" solution. This approach involves the use of cameras and artificial intelligence to swiftly assess the correct-ness, legibility, and presence of labels. In case of inaccuracies, a machine seam-lessly removes the product from the production line without causing a complete halt. Suntory PepsiCo relies on this Machine Vision System to uphold the high quality of their products. The manual inspection of each item by a human would be signifi-cantly more time-consuming, whereas AI-powered computers accomplish the task more swiftly and with fewer errors [7] .

4. Silicon wafers identify the root cause of microchip faults.

Silicon wafers, essential in semiconductor production for devices like phones, com-puters, and televisions, form the basis for creating minute chips, sometimes as small as 10 nanometers. Detecting manufacturing defects in these chips involves the use of electron microscopes, which, while precise, are slow. Despite the efficiency of opti-cal scans in identifying various faults on silicon wafers, the subsequent verification with an electron microscope is time-consuming. This delay is critical as even minor defects can lead to malfunctions in the chips, referred to as "killer" flaws. Applied Materials addresses this challenge through ExtractAI, a technology employing AI to identify these potentially detrimental flaws. The process involves utilizing a special-ized scanner to identify flaws on silicon wafers, which are then subjected to detailed examination using an electron microscope [8].

10.8 CONCLUSION

Advanced manufacturing technologies and rapid prototyping will empower individual consumers to order unique devices without incurring substantial expenses. The implementation of comprehensive simulation and virtual testing across the product lifecycle, facilitated by collaborative Virtual Factory (VF) systems, is poised to significantly reduce the costs and time associated with new product design and manufacturing process engineering. Advanced technologies such as human-machine interaction (HMI) devices and augmented reality (AR) will play a pivotal role in enhancing manufacturing plant safety and alleviating physical stress on an aging workforce.

Machine learning will be crucial for optimizing industrial processes, leading to reduced lead times and energy consumption. The integration of Cyber-Physical Systems and machine-to-machine (M2M) communication will enable the real-time collection and sharing of data from the shop floor, minimizing downtime and enhancing efficiency through highly effective predictive maintenance. Intelligent Manufacturing can manifest through various approaches, with key technical fields supporting intelligence in production.

S.No	Technology	Definition
1	Industrial Internet of Things (IIoT) [10]	This involves connecting people, machines, items, and ICT in real-time, intelligently, both horizontally and vertically, enabling the management of complex systems dynamically
2	Big Data Analysis [11]	The process of extracting knowledge from vast volumes of data to make informed decisions
3	Additive Manufacturing (3D Printing)	A process that serves as an alternative to traditional machining methods like lathes and milling machines
4	Cyber-Physical Systems (CPS) [12]	R encompasses augmented reality (AR), mixed reality (MR), and virtual reality (VR), aiming to bridge the gap between the physical and virtual worlds
5	Extended Reality (XR) [13]	The combination of augmented reality (AR), mixed reality (MR), and virtual reality (VR) that aims to connect the physical and virtual worlds
6	Cloud Services for Products [14]	Utilizing cloud computing in products to enhance their capabilities and offer associated services
7	Flexible Manufacturing	Digital automation of manufacturing processes using sensors like RFID or the development of reconfigurable manufacturing systems (RMS))

S.No	Technology	Definition
8	Virtual Mode Simulations/Analysis	Virtual model analysis using finite element analysis and computational fluid dynamics, in which models simulate attributes of implemented models
9	Integrated Engineering Systems [15]	To communicate information, IT support systems are incorporated into product development and manufacturing

REFERENCES

1. Chen, W., M. Nguyen, and P. Tai. "An intelligent manufacturingsystem for injection molding." *Proceedings of Engineering and Technology Innovation* 8.9 (2018). https://doi.org/10.1007/s00170-015-7683-0
2. Javaid, Mohd, et al. "Understanding the adoption of Industry 4.0 technologies in improving environmental sustainability." *Sustainable Operations and Computers* 3 (2022): 203–217.
3. Groumpos, Peter P. "The challenge of Intelligent Manufacturing Systems (IMS): the European IMS information event." (2021). https://doi.org/10.1007/bf00123677
4. V. Juyal, Nitin Pandey and Ravish Saggar, "Performance comparison of DTN multicasting routing algorithms-opportunities and challenges," *2017 International Conference on Intelligent Sustainable Systems (ICISS)*, Palladam, 2017, pp. 53–57. doi: 10.1109/ISS1.2017.8389238
5. Wang, Baicun, et al. "Smart manufacturing and intelligent manufacturing: A comparative review." *Engineering* 7.6 (2021): 738–757.
6. Annanth, V. Kishorre, M. Abinash, and Lokavarapu Bhaskara Rao. "Intelligent manufacturing in the context of industry 4.0: A case study of siemens industry." *Journal of Physics: Conference Series* 1969.1 (2021).
7. Jardim-Goncalves, Ricardo, David Romero, and Antonio Grilo. "Factories of the future: challenges and leading innovations in intelligent manufacturing." *International Journal of Computer Integrated Manufacturing* 30.1 (2017): 4–14.
8. Zahir, Shahirah Binti, et al. "Smart IoT monitoring system." *Journal of Physics: Conference Series* 1339.1 (2019).
9. Singh, Anamika, Akkas Ali, Balamurugan Balusamy, Vandana Sharma, Chapter 12 - Potential applications of digital twin technology in virtual factory. In: Rajesh Kumar Dhanaraj, Ali Kashif Bashir, Vani Rajasekar, Balamurugan Balusamy, Pooja Malik, *Digital Twin for Smart Manufacturing*, Academic Press, 2023, 221–241. https://doi.org/10.1016/B978-0-323-99205-3.00011-0.
10. A. Raj, V. Sharma, S. Rani, A. K. Shanu, A. Alkhayyat and R. D. Singh, "Modern Farming Using IoT-Enabled Sensors for the Improvement of Crop Selection," *2023 4th International Conference on Intelligent Engineering and Management (ICIEM)*, London, United Kingdom, 2023, pp. 1–7. https://doi.org/10.1109/ICIEM59379.2023.10167225.
11. A. Raj, V. Sharma, S. Rani, B. Balusamy, A. K. Shanu and A. Alkhayyat, "Revealing AI-Based Ed-Tech Tools Using Big Data," *2023 3rd International Conference on Innovative Practices in Technology and Management (ICIPTM)*, Uttar Pradesh, India, 2023, pp. 1–6. https://doi.org/10.1109/ICIPTM57143.2023.10118162.

12. Ali, Akkas, B. Balamurugan, and Vandana Sharma. "IoT and blockchain based intelligence security system for human detection using an improved ACO and heap algorithm." *2022 2nd International Conference on Advance Computing and Innovative Technologies in Engineering (ICACITE).* IEEE, 2022.

13. Srivathsan, K., Bharath, S., Malini, A. et al. "Extended virtual reality based memory enhancement model for autistic children using linear regression." *International Journal of System Assurance Engineering and Management* (2024). https://doi.org/10.1007/s13198-023-02231-5

14. Sharma, V., B. Balusamy, J. J. Thomas, and L. G. Atlas (Eds.). (2023). *Data Fabric Architectures: Web-Driven Applications.* Walter de Gruyter GmbH & Co KG.

15. V. Juyal, R. Saggar and N. Pandey, "Clique-based socially-aware intelligent message forwarding scheme for delay tolerant network," *International Journal of Communication Networks and Distributed Systems* 21(4) (2018): 547–559.

11 Case Studies of Next Generation AI for Intelligent Manufacturing

Gautam Samblani and Devershi Pallavi Bhatt

11.1 INTRODUCTION

Artificial Intelligence (AI) is becoming a part of almost every aspect of our everyday existence. Healthcare, geology, customer data analysis, self-driving cars, and even the arts are among its applications [1]. Its use is ever-changing, and its presence is all around us. However, there are still as many questions as there are answers regarding AI, including how to precisely define the technology and its various applications, such as support, augmentation, and autonomy.

The production industry is expected to see significant growth in the 2020s due to the continuous digital landscape transformation as the new decade gets underway. Artificial Intelligence is without a doubt the driving force behind this revolutionary voyage.

The manufacturing value chain could undergo a complete transformation owing to the potential of AI. Examples of this include direct automation, predictive maintenance, decreased downtime, continuous around-the-clock production, improved security measures, lower operating costs, increased efficiency, better quality control, and quicker decision-making. These are just a handful of the benefits companies can experience from successfully incorporating AI into their daily operations [2].

The good news for individuals who have been reluctant to embrace AI is that you are not alone in this path, and it is not too late to start. Despite the fact that the manufacturing sector is frequently seen as a leader in the adoption of new technology, just 9% of manufacturing executives from 26 countries surveyed by PwC in 2018 said they have integrated AI into their operational decision-making processes. This low adoption rate is probably caused by the fact that putting AI into practice is a difficult, resource-intensive task that requires a thorough and methodical strategy to be proven successful.

Although adopting Artificial Intelligence may seem intimidating, it is now generally accepted that, within the next five years, AI will have a major impact on every industry. Businesses and sectors that are ahead of their rivals in this technology transformation will reap substantial benefits. Those that disobey run the danger of losing their edge over competitors. Because technology is developing so quickly, businesses that delay implementing AI run the risk of slipping behind before they know it.

DOI: 10.1201/9781032630748-11

217

Artificial Intelligence is at the heart of the concept of "Industry 4.0," which prioritizes higher automation in manufacturing environments as well as the generation and
exchange of vast volumes of data there. AI has to be used by companies, along with
Machine Learning (ML), to optimize this big data that is generated by manufacturing equipment. There are several benefits associated with AI-based data analytics
and usage for manufacturing process optimization such as cost savings, better safety
protocols, increased supply chain efficiency, among others [3].

This chapter discusses how Industry 4.0 heavily relies on Artificial Intelligence
to automate production processes and use data effectively. The advent of Industry
4.0 has led to massive amounts of information being produced in their facilities
by manufacturers. This includes information from sensors, devices, and operations
amongst others. The data created within a manufacturing environment can be a valuable asset with substantial advantages when it is properly analyzed and put to use.
Artificial Intelligence and Machine Learning: AI and ML techniques are needed
for comprehending this data [4]. They are able to recognize trends, anticipate future
events, and enhance production procedures. Using AI in manufacturing can lead to
a number of advantages, such as reduced costs, increased safety, better supply chain
performance, and general process optimization.

11.2   THE ROLE OF AI IN SMART MANUFACTURING

Automation revolutionized production lines in the twentieth century, but AI is ushering in a new era of smart manufacturing by elevating manufacturing to new heights.
Whenever we discuss smart manufacturing, the topic invariably revolves around the
application of sensors, robots, IoT, machine-to-machine communication, and more.
But these detached, indifferent components don't contribute to more intelligent
manufacturing [5]. Robots, for instance, are limited to what they can do by their
programming. The amount of data generated by sensors is limited. An intelligent
system that can handle the amount of data produced by the Smart Manufacturing
Network is what these systems require in order to give companies optimal processes,
which boost productivity, capacity, and efficiency while enabling factories to make
better business decisions and innovate more quickly [6].

Today's factories are fundamentally a network of processes made up of information flows and tasks due to the development of automation in the manufacturing
sector. In order for a manufacturing facility to become intelligent, these procedures
need to run smoothly and continuously. Siemens senior vice president of Internet of
Things (IoT) and go-to-market strategy, Jagannath Rao stated, "AI enables various
aspects of operations ... to predict outages to avoid them." downtime; carry out the
best production schedules depending on the orders that are currently in place; and
make forecasts, inventories, and delivery schedules ... AI enables manufacturers to
employ advanced robotics for improved precision and quality, as well as to plan
their own logistics to ensure a smooth delivery process. AI may accomplish this by
introducing cutting-edge automation to production procedures, such as scheduling,
predictive maintenance, and planning.

A smart factory is a networked system of manufacturing facilities where information is connected to obtain deep and instantaneous insights from data from supply chains, production lines, design teams, and quality control. By applying AI to the data collected from the linked network, these actionable insights help make smarter decisions [7].

11.2.1   AI AND SMART MACHINES

It is true that computers can only perform what we tell them to do in the absence of flexible algorithms, according to Michael Mendelson, a program developer at NVIDIA's Deep Learning Institute (DLI). (The NVIDIA Deep Learning Institute offers resources for diverse learning needs—from learning materials to self-paced and live training to educator programs.) It becomes challenging to translate cognitive tasks into rule-based instructions in a production setting. Manufacturers can resolve this paradox and enable AI to allow manufacturing robots to perform cognitive activities more adeptly, communicate with one other more effectively than they would with human partners, and obey their commands.

By creating a common language that both humans and machines can understand, AI is what makes these concepts useful and usable in the manufacturing ecosystem. Examples of AI applications include industrial vision, which provides precise analysis of quality; and cobot development, which involves teaching a robot to detect and avoid danger or disturbance in its environment [8,9].

11.2.2   AI AND THE SUPPLY CHAIN

If AI is the cognitive ability that makes machines smarter and makes it easier for humans and robots to work together, then AI's contribution to smart manufacturing goes much beyond. Everything is optimized and waste is reduced in the smart factory. A smart factory's manufacturing supply chain functions similarly.

Manufacturers can use AI to identify demand patterns across time periods, regions, and socioeconomic groups. They can also consider the influence of macroeconomic and political changes, weather, business cycles, and other factors on the supply chain. This makes it possible to estimate market demand more accurately, which has an impact on hiring practices, payroll policies, the acquisition of raw materials, inventory management, energy usage, and equipment maintenance. Additionally, it provides manufacturers with hints to forecast demand and produce products to fill the pipeline [10].

11.2.3   AI AND MAINTENANCE

Vast networks of connected devices generate a wealth of data in the smart factory. AI uses data from this connected network to enable predictive maintenance. The industrial sector must improve predictive maintenance capabilities to anticipate maintenance events, extend predictive maintenance, and enjoy the economic benefits that smart factories deliver machine lifecycles, improve production timetables, and individually adjust each department's production and maintenance schedules.

Predictive analytics further benefits from Artificial Intelligence by assisting in the avoidance of unanticipated and undetected equipment faults. When a flaw is found or a new part is required, AI allows the program to order it and carry on with the process uninterrupted. AI also makes it possible for manufacturers to build "digital twins" that track performance data continually and generate predictive analytics that schedule services according to need rather than availability. Factories that possess this capacity are able to manage issues related to unscheduled maintenance, low output, and low throughput [11].

11.2.4 AI Makes QC Smarter in the Smart Factory

Manufacturers can provide adaptive machine control and boost machine productivity with the help of AI and better monitoring capabilities. On the same machine, producers may produce more exact items with improved system planning and monitoring.

AI also makes automated quality control possible in ecosystems for smart manufacturing. Quality control is crucial since complicated products are becoming more and more varied. Manufacturers may now quickly identify mistakes, detect anomalies and pinpoint the primary cause of many device failures in seconds as opposed to hours by utilizing Artificial Intelligence. This aids producers in proactively resolving problems before they result in expensive delays.

The primary advantages of AI for smart manufacturing include increased productivity, cost control, shorter time to market, process optimization, and performance improvement. Additional benefits of AI in the smart manufacturing ecosystem include machines' capacity for self-learning through experience and their availability around-the-clock. With a deeper grasp of predictive trends, AI has the potential to revolutionize how humans and machines interact and produce better results. Critical elements like security and safety can be optimized with AI assisting operational choices to make the production floor a safer place for people.

Smart manufacturing is undoubtedly a step in the direction of manufacturing excellence. But using AI is the only way to turn these clever efforts' potential into real outcomes [12,13] (Table 11.1).

11.3 BENEFITS AND MOTIVES FOR ADOPTING AI IN MANUFACTURING

One benefit is garnering lots of information from Internet of Things via smart factories. Manufacturers are using AI methods such as deep neural networks or Machine Learning (ML) to analyze these pieces and enhance the decision-making procedures.

Here are the main uses of AI in the manufacturing sector:

- Machine Learning: Programs that learn from patterns in data instead of being explicitly programmed.

TABLE 11.1

Key Statistics Related to AI in Manufacturing

Statistic	Description
Market Growth (2022)	Approximately 2.3 billion dollars
Projected Market Growth (2027)	Expected to reach 16.3 billion dollars, Increasing at a compound annual growth rate (CAGR) of 47.9 % annually.
Implementation of AI based on Function	Maintenance (29%) and Quality (27%)
Industrial Robots Market (2022)	Over 680,000 units
Predictive Maintenance Impact	Reduces machine maintenance costs by up to 25%, leading to 70% fewer breakdowns
Supply Chain Optimization Benefits	Logistics cost improvement by 15%, inventory levels by up to 35%, and service levels by up to 65%
Error Detection Accuracy (Quality Control)	Up to 90% improvement in overall product quality
Reduction in Machine Downtime (Predictive Maintenance)	35% to 45% reduction in machine downtime
AI-Enhanced Production Rates	Up to 20% increase in production rates
AI in Generative Design	Accelerates design processes and reduces material waste
AI in Inventory Management and Demand Forecasting	Reduces inventory issues and improves demand forecasting
AI in Factory Layout Optimization	Identifies inefficiencies, eliminates bottlenecks, and improves performance

- Computer Vision: A more sophisticated type of Machine Learning meant to help in image and video interpretation.
- Autonomous Systems [14]: Intelligent cars or robots that answer questions by themselves.

Artificial Intelligence and manufacturing together have transformed industrial processes and stimulated greater innovation in the manufacturing industry. Through this partnership, factories and other industries may leverage AI's promise to improve operations, make data-driven decisions, and create intelligent, adaptable systems.

Let's examine some figures that illustrate how AI is affecting manufacturing:

- Market Growth: In 2022, the Artificial Intelligence market in manufacturing is expected to reach a valuation of 2.3 billion dollars. But with a strong 47.9 % annual growth rate, it should reach an astounding 16.3 billion dollars by 2027.
- AI Implementation by Function: Maintenance (29%) and quality control (27%) are the two primary crucial tasks for which manufacturers employ AI.

- Automation Revolution: There is a major automation shift taking place in the manufacturing sector. Thanks to developments in AI and robotics technology, the number of industrial robots sold worldwide is predicted to surpass 680,000 units by 2022. AI-enabled robots are becoming more and more popular. By 2031, this market might be valued at $150 billion.
- Better Quality: There are notable gains when AI and intelligent picture recognition are used for quality assessment. AI can achieve up to 90% accuracy rates in error detection, which improves total product quality. This can lead to a 50% boost in productivity for factories.
- Predictive Maintenance: AI-powered predictive maintenance can reduce machine maintenance expenses by up to 25%, which will lead to a 70% decrease in breakdowns. This technology optimizes maintenance plans and helps prevent expensive breakdowns.
- Supply Chain Optimization: Factories are significantly impacted by supply chain management driven by AI. AI early adopters should expect up to a 15% reduction in logistical costs, a 35% decrease in inventory levels, and a 65% boost in service levels.

11.3.1   KEY BENEFITS OF AI IN MANUFACTURING

Incorporating robotics and Artificial Intelligence into production has many advantages, some of which are listed below:

Predictive maintenance: AI-driven predictive maintenance makes use of Machine Learning, sensor data from the machine (such as vibration, motion, and temperature), and external data from the weather. By anticipating possible machine faults, this technique lowers maintenance costs, minimizes unscheduled downtime, and increases the lifespan of the unit. Data from the US Department of Energy show that predictive maintenance can save downtime for machines by 35 to 45 %.

Intelligent quality control (IQC): AI offers a thorough insight into manufacturing facilities, assembly lines, and warehouses, facilitating the identification of quality problems as well as the elimination of waste and enhancement of processes. AI has the potential to improve quality by up to 35% while helping manufacturers raise production rates by up to 20%. AI-powered sensors are extremely good at picking up on even the smallest defects that a human eye might miss. Productivity and the rate at which products pass quality control both rise as a result. Additionally, regular operations are automated by AI, which increases accuracy and eliminates the need for labor-intensive and prone-to-error human inspections.

AI can be instrumental in activities involving complex mathematics because it is good at simplifying intricate mathematical calculations and coding. AI turns these duties into automated tasks. Hence, making them handy tools for engineers to speed up their work. In so doing, the workers are saved from engaging in repetitive mechanical functions that consume time and thus get an opportunity to become more creative, which enhances job satisfaction and makes people more productive. This technology increases worker productivity by providing fast access to crucial data. With reports, manufacturers can make order predictions while engineers may discover relevant materials for a product quickly.

Generative design is a major advantage of AI in manufacturing. The use of AI algorithms to explore multiple designs for various products and components greatly supports efficiency and creativity in the process of designing them.

This includes such improvements as:

- Innovative design: this can provide unique solutions that were unimaginable to humans.
- Efficiency: the ability of manufacturers to provide design alternatives that have not come to the human designer's mind.
- Less waste of materials: it is possible to reduce material usage and contribute to sustainable programs by creating lightweight and efficient structures.
- Improved demand forecasting and inventory management: there are AI-powered predictive models for demand, which can be used in coordinating inventory between multiple locations as well as managing inventory within complex global supply chains.
- AI-powered forecasts: according to McKinsey Digital, these forecasts can cut supply chain errors by as much as 50% . With the help of solid calculations made by AI, the forecasts maintain a true balance of stock, project demand manage inventory over intricate international supply chains, and coordinate inventory levels across locations.

Customize the factory layout: it can be challenging to create an efficient factory floor due to the many factors, and the layout needs to change as the product does. AI solutions can help find floor plan inefficiencies, remove bottlenecks, and boost productivity. Manufacturers are able to undertake rapid testing without any interruptions and swiftly assess industrial traffic conditions after making modifications.

11.4   EXAMPLES OF AI APPLICATIONS IN THE MANUFACTURING INDUSTRY

In the past, factories with employees were only seen in sci-fi stories. Today manufacturers have brought this concept to life by incorporating AI into their operations. AI has significantly impacted the manufacturing industry leading to changes in how products are made. Below are some key examples of how AI is being used in this sector.

Collaborative Robotics (Cobots) [15]

Collaborative robots, also known as "cobots" often work alongside employees in a supporting role. Unlike robots designed for tasks, cobots have the ability to learn various functions.

They are capable of recognizing and avoiding obstacles, which promotes a co-existence with their human counterparts. Manufacturers commonly use cobots in labor and assembly line environments. For example, in an automobile factory cobots can handle the lifting of car parts while a human worker secures them in place.

These collaborative robots are skilled at locating and retrieving items, making them valuable assets in warehouses. Equipped with intelligence, computer vision, and learning capabilities, cobots excel at performing tasks that require intricate assembly and quality checks.

Robotic Process Automation (RPA) for Streamlining Paperwork:
AI and ML technologies have greatly enhanced the manufacturing sector in managing paperwork by utilizing Robotic Process Automation (RPA). Manufacturing plants often deal with a large amount of paperwork including orders, invoices, and reports. Handling documents manually is time consuming and prone to errors which can lead to challenges.

Nevertheless, incorporating AI-driven bots allows for the automation of these back-end processes. Smart robots skilled in analyzing documents can interpret, categorize, and accurately input information into the systems. For example, a car manufacturer could utilize robots to process supplier invoices.

These robots excel at understanding details validating their correctness and entering the data into systems without the need for manual input.

Revolutionizing New Product Development
The use of AI has brought about a change in how manufacturing facilities approach the development of products. This shift allows companies to be more innovative and introduce products effectively. One key benefit of AI is its ability to quickly analyze sets of data.

Smart algorithms extract insights from consumer preferences and competitor actions aiding in making informed decisions about product offerings. Additionally, AI plays a role in the design stage by working alongside engineers to generate a variety of design options based on requirements.

This speeds up the design process leading to more efficient product development. By incorporating AI, companies can operate with flexibility, creating products that meet market needs, closely enhancing competitiveness, and bringing innovative ideas to life swiftly.

11.5   APPLICATIONS OF AI IN INTELLIGENT MANUFACTURING

Manufacturing plays a role in a country's economy, impacting life and contributing to national security. The merging of manufacturing technology with information technology, intelligent systems and specialized knowledge is gradually reshaping how manufacturing operates and its surrounding ecosystems.

In today's world, the widespread use of the Internet, sensors, big data, e-commerce advancements and information communities has intertwined data and knowledge with society, physical spaces, and the digital realm. This shift in the information landscape is influencing the development of intelligence, propelling it into a new phase of growth.

This evolution is characterized by the introduction of technologies that are driving AI forward. AI 2.0 brings capabilities like data-driven recognition for focused deep learning, swarm intelligence based on internet connectivity, human-machine collaboration centered around technology, enhanced augmented intelligence, and the rise of cross-media cognitive processes [16].

The increasing need for AI is being fueled by progress in sectors such as cities, healthcare, transportation, logistics, robotics, autonomous vehicles, smartphones, and more.

These progressions are transforming our communities, driving innovations in AI technologies and their uses.

The integration of AI into intelligent manufacturing can be assessed from three dimensions:

- Applied Technology:

This dimension evaluates the level and capacities of infrastructure development, individual application domains, collaborative application scenarios, and the maturation of business implementations.

- Industrial Advancement Assessment:

This involves assessing intelligent products (those capable of autonomous and intelligent operation), intelligent interconnected products (products forming ecological networks), intelligent industrial software, and the intelligent design, production, and management processes. This includes hardware development, support, security, commissioning, and the operation of intelligent manufacturing systems at multiple levels, spanning intelligent manufacturing units, workshops, factories, and entire industries.

- Application Effectiveness:

Evaluation within this aspect focuses on changes in competitiveness and associated alterations in societal and economic benefits. It involves measuring the direct or indirect impact of intelligent manufacturing systems on performance enhancements and economic benefits [8].

11.5.1 CASE STUDY: BMW's APPLICATION OF AI ALGORITHMS

The BMW Group has implemented digitalization efforts at its factories utilizing an approach that involves end-to-end platforms, integrated technologies, autonomous robotic systems and collaborative interactions between humans and robots. This collective endeavor aims to boost efficiency and support model customization across the BMW Group's production networks. With a sales figure of 2.5 million vehicles, the company fulfills 99 % of customer orders for multiple car models, each with unique specifications. However, managing the influx of millions of parts from suppliers worldwide poses logistical challenges for BMW Group's manufacturing facilities.

To address these obstacles, BMW Group has integrated robotics and Artificial Intelligence into its operations resulting in noteworthy achievements within its manufacturing plants.

The BMW Group achieves a daily production output of 10,000 vehicles across a variety of 40 different vehicle models. Each model offers an average of 100 configuration options leading to a range of vehicle configurations.

The logistics department at the BMW Group efficiently processes an intake of 30 million parts sourced from 1,800 suppliers spread across 31 factories.

These components, totaling 230,000 part numbers, are carefully arranged in compartments. It is impressive to note that the BMW Group manufactures 10,000 custom vehicles daily, demonstrating their efficiency in meeting customer needs.

The BMW Group's production line operates at a pace with a vehicle completing assembly every 56 seconds. This continuous production line is flexible enough to manufacture up to 10 vehicle models in any sequence, ensuring delivery of parts to their designated locations at the right time.

Device Chronicle conducted interviews with experts involved in this project such as Michael Schmidt, General Manager of Shop Floor Digitalization Production Systems at the BMW Group, and Carolin Richter from the Innovation and Digitalization Group of the BMW Group.

11.5.1.1 Formulating the Platform Strategy

Overseeing operations both in production and shop floors, Mr. Schmidt plays a role in leading the production strategy at BMW Group. A recent interview of him is highly centered on the importance of cloud-to-edge platforms that drives his strategic approach to integrate Operational Technology (OT) and Information Technology (IT). In addition, he explains how real-time communication connects systems and production units to higher-level systems.

Schmidt notes, "Data processing at the edge can only be possible by ensuring seamless real-time communication between production units, robots and higher-level systems." This is indicative of a change in manufacturing technology as there is an emergence of flexible production cells.

Moreover, Schmidt opines, "Hardware systems know hardware while applications embody software. Comparatively, software has more flexibility than hardware. Moreover, this report discusses the goals and challenges faced by BMW Group along this journey.

"To date our primary objective has been about enhancing digitalization efforts across existing production units with a special focus on bettering product quality and increasing productivity. All these efforts are guided by one overarching goal."

Over the years we have been focused on introducing new products onto our assembly line. These include cameras, graphics processing units (GPUs), and various other small, embedded devices.

Moreover, there are two goals that Michael Schmidt has highlighted:

It is important to have efficient central management for several edge devices due to reasons of cybersecurity. Firmware should be kept up-to-date, application patches done in good time and any replacements when necessary. The significance of this is underscored by Schmidt as it allows maintenance personnel to replace devices locally while centrally managing software configurations for these devices. For instance, an app store is very important for making applications accessible via such devices.

He continues to say that edge devices also connect with other devices. This aims at placing the edge close to a production facility within a data center so that it can work as a self-governing entity [9].

11.5.2 ALTERNATIVE APPLICATIONS OF AI IN INTELLIGENT MANUFACTURING

11.5.2.1 AI-Powered Process Automation

The field of intelligence has seen the rise of process mining tools that can automatically identify and fix bottlenecks in manufacturing processes. These advanced tools also allow for comparing the performance of manufacturing facilities across regions. This gives manufacturers insights into how they are operating, helping them standardize and optimize their workflows to improve manufacturing processes.

Another interesting use case is in Robotic Process Automation (RPA) where autonomous robots handle tasks in shop floor settings. Humans step in when the robot faces exceptions or deviations from the process. Robots use computer vision to examine and filter processes without needing human intervention.

Process automation, enabled by AI, yields additional benefits, including:

- Reduction in cycle time
- Amplified production output
- Enhanced precision in operations
- Improved workplace safety standards
- Elevated employee morale and productivity levels

The incorporation of AI to automate processes in the semi-conductor industry, according to McKinsey, can yield productivity improvements of up to 30 %.

This integration also helps reduce scrap rates and testing expenses, further enhancing the efficiency and economic viability of the industry (Figure 11.1).

11.5.2.2 AI in warehouse management

AI can automate some aspects of warehouse operations. By collecting real-time data, manufacturers can continuously monitor their warehouses and better plan logistics. Demand forecasting can help manufacturers take steps to stock their warehouses in advance and meet customer demand without incurring large transportation costs. Warehouse robots can track, lift, move and sort items, delegating the most strategic tasks to humans and reducing workplace accidents. Automated quality control and inventory can reduce warehouse management costs, improve productivity, and

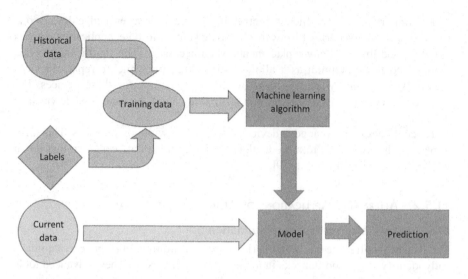

FIGURE 11.1   AI automation.

reduce labor requirements. Therefore, manufacturers can increase their sales and profits.

11.5.2.3   Integration of Artificial Intelligence and IoT (AIoT) in Manufacturing

The concept of IoT, denoting interconnected smart devices equipped with sensors capable of generating substantial volumes of real-time operational data, has found its industrial counterpart known as IIoT, or the Industrial Internet of Things, within the manufacturing landscape. When coupled with the power of AI, this amalgamation, often referred to as AIoT, unlocks the potential for manufacturing processes to attain heightened levels of precision and productivity.

Within the sphere of IIoT, several critical use cases materialize, including (Figure 11.2):

- Adoption of wearable devices, such as smart glasses, designed to furnish hands-free instructions and real-time situational awareness to workers.
- Continuous monitoring of equipment performance, energy consumption patterns, ambient temperature variations, and the detection of hazardous gases, thereby bolstering workplace safety.
- Implementation of intelligent lighting and HVAC controls to optimize energy consumption within manufacturing facilities.
- Leveraging industrial analytics to harness data emanating from advanced equipment positioned on the manufacturing floor.

These innovative applications at the intersection of AI and IIoT serve as pivotal contributors to the ongoing optimization of manufacturing operations, fostering environments that are not only more efficient but also safer for the workforce [17,18].

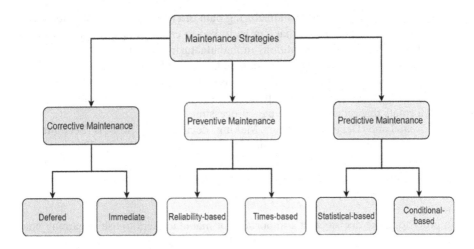

FIGURE 11.2   AI maintenance system.

11.6   CHALLENGES AND BARRIERS TO AI IMPLEMENTATION IN MANUFACTURING

Manufacturers are leading the way in incorporating Artificial Intelligence (AI) technology into their operations. They are using AI-powered analytics to improve efficiency, product quality, and workplace safety. However, they face challenges when trying to meet deadlines for bringing products to market quickly, creating more complex products and complying with strict quality regulations and standards. Despite their efforts, many manufacturing companies must overcome obstacles that impede their transformation and AI initiatives.

1. Scarcity of AI Talent: Skilled data scientists and AI experts are in short supply. Successful AI projects require a team collaboration involving data scientists, Machine Learning engineers, software architects, business intelligence analysts, and subject matter experts. This demand is especially evident in the manufacturing industry, which is often seen as unexciting by emerging data scientists. Adding to this challenge is the labor shortage in manufacturing due to the retirement of the baby boomer generation. AI automation and AutoML 2.0 are emerging as technologies that can help bridge this skills gap and accelerate transformation within manufacturing.
2. Technology Infrastructure and Interoperability: Manufacturing facilities typically have a variety of machines, tools, and production systems that run on technologies that may not always work well together. In some cases, these machines may be running on outdated software that doesn't work well with the rest of the system. Without frameworks and common standards, in-place plant engineers need to figure out the ways to connect their machinery and systems as well as decide which sensors or transducers to use.
3. Data Quality: Having access to meaningful and high-quality data is crucial for the success of AI projects. It can be tough to come by in the

manufacturing sector. Manufacturing data often contains biases, outdated information, and errors that can be linked to causes. For example, sensor data collected under conditions in manufacturing facilities might produce results due to significant temperature changes, noise levels, and vibrations. Historically, factories have had systems that don't work together seamlessly, leading to operational data spread across different databases with unique formats that require extensive preprocessing.

4. Real-Time Decision-Making: Making decisions in time is becoming increasingly important in manufacturing applications for quality control and meeting customer delivery deadlines. Often decisions need to be made within seconds to identify problems before they cause downtimes, mistakes, or security issues.

In order to make decisions, it's important for manufacturers to use streaming analytics and real-time prediction services. This helps them react promptly and avoid outcomes.

Edge computing plays a role in manufacturing by allowing companies to process data locally and filter information which decreases the amount of data sent to central servers whether they are on site or in the cloud. Moreover, modern manufacturing focuses on using data from machines, processes, and systems to make real-time adjustments during production processes. This careful monitoring and control of assets and manufacturing operations involve data usage.

Rely on Machine Learning for making informed decisions based on insights derived from data. The capability to deploy models to edge devices such as machines, local gateways, or servers is crucial for supporting manufacturing applications.

5. Building trust: Ensuring transparency is essential in the adoption of AI technology by manufacturers. The complexity of AI technology combined with uncertainties about its capabilities pose a challenge to AI implementation. People without a background in data science find it difficult to understand how data science works and how predictive modeling functions, leading them to question the algorithms that underlie AI technology. Improved transparency in the AI process is crucial to providing insights into data processing methods, algorithm choices, and reasoning behind predictions.

Manufacturing organizations can build trust in AI models. Gain business confidence by understanding how predictive models work and why they make predictions. AI models, newer data science approaches offer more transparency throughout the AI process. This includes a look at how raw data is transformed into Machine Learning inputs (such as feature engineering) and how ML models use various features to make predictions. By providing insights into how predictive models operate and the reasons behind their predictions, manufacturing organizations can establish trust in these models. Appreciate the value they bring to the business in terms of increasing profits, helping in decision making, predicting future trends and many more.

TABLE 11.2

Challenges and Barriers to Implementation

Challenge or Barrier	Description
Lack of AI Talent	Difficulty in finding experienced data scientists and AI experts. The sector faces a potential labor shortage in the coming years.
Technology Infrastructure and Interoperability	Varied and incompatible technologies and systems in manufacturing facilities can hinder integration and data flow.
Data Quality	Poor data quality, including bias, errors, and inaccuracies, can impact the effectiveness of AI initiatives.
Real-Time Decision- Making	In manufacturing, rapid decision-making is crucial, and AI solutions are needed to enable real-time actions and predictions.
Edge Computing	The ability to deploy predictive models and AI solutions at the edge, closer to production, is essential for smart manufacturing.
Trust and Transparency	Building confidence in AI models and algorithms among non-experts is challenging but crucial for widespread adoption.

Despite the challenges faced, AI adoption remains the pathway for manufacturers to enhance automation, reduce costs, boost efficiency, and tackle emerging challenges more effectively [11] (Table 11.2).

11.7   FUTURE TRENDS IN AI FOR INTELLIGENT MANUFACTURING

The manufacturing sector faces obstacles such as upkeep of machinery, inefficiencies in operations, and flaws in products. Thankfully, the integration of knowledge with AI and data-centric technology known as Industry 4.0 is transforming the industry. Through the advent of the Internet of Things (IoT) and automated factories, a greater volume of data is produced daily. AI plays a role in deciphering this data to derive meaningful insights. According to GP Bullhound, the manufacturing field generates 1,812 petabytes (PB) of data annually, surpassing sectors like BFSI (Banking, Financial Services and Insurance), retail, and media. Manufacturers are leveraging AI tools such as Machine Learning, deep learning, and natural language processing to analyze data for decision-making [21].

AI has applications throughout the production process. From sourcing materials and manufacturing to distributing products. One key use of AI is maintenance, which enables manufacturers to anticipate and prevent machinery breakdowns proactively, thereby reducing production downtime. AI in manufacturing also shows potential in areas such as forecasting demand, ensuring quality control by conducting inspections, and automating warehouse processes. The concept of "Industry 4.0" revolves around AI and ML to help organizations harness the value derived from datasets generated by production equipment.

The use of AI in handling this data can lead to cost reductions, improved safety measures, increased efficiency in the supply chain, and various other advantages. In the field of AI within the manufacturing sector, subtechnologies like Machine

Learning, deep learning, language processing, and machine vision play roles across a range of processes. When it comes to ML technology in manufacturing, it can be classified into two types; supervised and unsupervised Machine Learning. Supervised ML involves utilizing AI to recognize patterns in datasets with known outcomes. This is particularly useful for predicting the remaining life of machinery and assessing the probability of failure for equipment. On the other hand, unsupervised Machine Learning focuses on identifying patterns in datasets where outcomes are not predetermined. For instance, engineers can leverage ML technology to identify anomalies and faulty components within a production line.

According to a Market Research Futures report, "the global market for AI applications in manufacturing was valued at $2.45 billion in 2022. [This] is estimated to increase from $3.61 billion in 2023 to $53.69 billion by 2030 with a growth rate of 47.1% during the period from 2023, to 2030" [22]. The use of AI is currently not widespread. It is expected to increase in the future due to advancements in Industry 4.0 regulations, the rise of automation, and higher investments in AI by tech companies. There are opportunities for those involved in both developed and emerging markets. Key players in the AI manufacturing sector include countries such as the United States, China, Germany, the UK, France, Italy, Japan, India, and South Korea.

Looking forward we might see automated factories where product designs are created with little to no involvement as AI technology continues to evolve. The onset of Industry 5.0 signals an era where AI works alongside capabilities. To keep up with this progress and drive AI adoption further into manufacturing, markets worldwide will require innovation through ideas, technological synchronization, unique applications, and cutting-edge advancements [19,20] (Table 11.3).

TABLE 11.3

Future Trends in AI for Intelligent Manufacturing

Future Trends in AI for Intelligent Manufacturing

1. Increased Adoption of Industry 4.0
2. Rise in Automation and IoT Integration
3. Enhanced Predictive Maintenance
4. AI-Driven Quality Assurance
5. Expansion of AI in Product Distribution
6. Advancements in Machine Learning Technologies
7. Growth in Edge Computing for Manufacturing
8. Emergence of Industry 5.0 with AI and Humans
9. Ongoing Innovation and Transformation
10. Global Market Expansion and Opportunities

11.8 CONCLUSION

Consequently, utilization of Artificial Intelligence in the manufacturing industry known as Industry 4.0 has led to an era of effectiveness, capability, and novelty to drive up productivity. AI offers benefits like maintenance, improved quality control, increased productivity, and enhanced inventory management and optimizing factory layouts for maximum results. These advantages are driving the market growth of AI in manufacturing with expectations of crossing milestones in the coming years.

Manufacturers are using AI technologies such as Machine Learning, deep learning, natural language processing, and machine vision to interpret huge amounts of data generated by production equipment. This gives way to a data-driven approach that enhances decision-making processes, reduces system downtime, and ensures compliance for efficient operations.

On the other hand, there are challenges that come with incorporating AI into manufacturing practices. Industry faces challenges including lack of experts in AI infrastructure complexities related to AI technology implementation, difficulties associated with real-time decision-making requirements due to data errors, as well as deploying edge-based AI models. Additionally, trustworthiness and transparency must be evident for all these models to be accepted.

In summary, therefore, it is clear that AI will continue revolutionizing and renewing the manufacturing sector into the future. Industry 5.0, the next stage in the evolution of manufacturing, is expected to bring about stronger collaboration between humans and AI, encouraging innovation and reshaping the manufacturing sector.

In this changing landscape, manufacturers need to be flexible and open to concepts and at the forefront of AI and automation to fully leverage the benefits of smart manufacturing. As AI technology progresses, it will lead to enhancements in efficiency, performance, and competitiveness, positioning manufacturing as a pivotal force in achieving manufacturing excellence.

REFERENCES

1. Atieh, Anas Mahmoud, Kavian Omar Cooke, and Oleksiy Osiyevskyy. "The role of intelligent manufacturing systems in the implementation of Industry 4.0 by small and medium enterprises in developing countries." *Engineering Reports* 5.3 (2023): e12578.
2. Plathottam, Siby Jose, et al. "A review of artificial intelligence applications in manufacturing operations." *Journal of Advanced Manufacturing and Processing* 5.3 (2023): e10159.
3. Zhang, Chao, et al. "Towards new-generation human-centric smart manufacturing in Industry 5.0: A systematic review." *Advanced Engineering Informatics* 57 (2023): 102121.
4. Nguyen, Huu Du, and Kim Phuc Tran. "Artificial intelligence for smart manufacturing in industry 5.0: Methods, applications, and challenges." *Artificial Intelligence for Smart Manufacturing: Methods, Applications, and Challenges* (2023): 5–33.
5. "Impact of Industry 4.0 on Supply Chains and Its Benefits," Website, 2019. [Online]. Available: https://imaginovation.net/blog/ai-in-manufacturing/

6. Ochoa, William, Felix Larrinaga, and Alain Pérez. "Context-aware workflow management for smart manufacturing: A literature review of semantic web-based approaches." *Future Generation Computer Systems* (2023).

7. Ryalat, Mutaz, Hisham ElMoaqet, and Marwa AlFaouri. "Design of a smart factory based on cyber-physical systems and internet of things towards industry 4.0." *Applied Sciences* 13.4 (2023): 2156.

8. Lei, Yu, et al. "Immersive virtual reality application for intelligent manufacturing: Applications and art design." *Mathematical Biosciences and Engineering* 20.3 (2023): 4353–4387.

9. Kirschbaum, Julius, Tim Posselt, and Kathrin M. Möslein. "Designing innovation ecosystems with functional roles: the case of industrial, intelligent manufacturing." *Handbook on Digital Platforms and Business Ecosystems in Manufacturing*. Edward Elgar Publishing, 2024. 106–125.

10. Li, Chengxi, et al. "Deep reinforcement learning in smart manufacturing: A review and prospects." *CIRP Journal of Manufacturing Science and Technology* 40 (2023): 75–101.

11. Singh, Amrinder, et al. "Smart manufacturing systems: a futuristics roadmap towards application of industry 4.0 technologies." *International Journal of Computer Integrated Manufacturing* 36.3 (2023): 411–428.

12. Yang, Zhihong, and Yang Shen. "The impact of intelligent manufacturing on industrial green total factor productivity and its multiple mechanisms." *Frontiers in Environmental Science* 10 (2023): 1058664.

13. P. Ajmani, V. Sharma, S. Sharma, A. Alkhayyat, T. Seetharaman and Z. Boulouard, "Impact of AI in financial technology- A Comprehensive Study and Analysis," *2023 6th International Conference on Contemporary Computing and Informatics (IC3I)*, Gautam Buddha Nagar, India, 2023, pp. 985–991, doi: 10.1109/IC3I59117.2023.10398111.

14. Pandey Anamika; Balamurugan Balusamy; Vandana Sharma, "Introduction," in *Disruptive Artificial Intelligence and Sustainable Human Resource Management: Impacts and Innovations – The Future of HR*, River Publishers, 2023, pp. 1–14.

15. A. Raj, A. Kumar, V. Sharma, S. Rani, A. K. Shanu and H. K. Bhardwaj, "Decipherable for Artificial Intelligence in Medicare: A Review," *2023 4th International Conference on Intelligent Engineering and Management (ICIEM)*, London, United Kingdom, 2023, pp. 1–6, doi: 10.1109/ICIEM59379.2023.10165690.

16. N. Manikandan, D. Rubey, S. Murali and V. Sharma, "Performance Analysis of DGA-Driven Botnets using Artificial Neural networks," *2022 10th International Conference on Reliability, Infocom Technologies and Optimization (Trends and Future Directions) (ICRITO)*, 2022, pp. 1–6, doi: 10.1109/ICRITO56286.2022.9965044.

17. V. Juyal, R. Saggar and N. Pandey, "Clique-based socially-aware intelligent message forwarding scheme for delay tolerant network," *International Journal of Communication Networks and Distributed Systems* 21.4 (2018): 547–559.

18. V. Juyal, Nitin Pandey and Ravish Saggar, "Performance comparison of DTN multicasting routing algorithms-opportunities and challenges," *2017 International Conference on Intelligent Sustainable Systems (ICISS)*, Palladam, 2017, pp. 53–57. doi: 10.1109/ISS1.2017.8389238.

19. V. Sharma, B. Balusamy, J. J. Thomas, and L. G. Atlas. (Eds.). *Data Fabric Architectures: Web-Driven Applications*. Walter de Gruyter GmbH & Co KG, 2023.

20. K. Durgalakshmi, P. Anbarasu, V. Karpagam, A. Venkatesh, B. Kannapiran and V. Sharma, "Utilization of reduced switch components with different topologies in multi-level inverter for renewable energy applications-a detailed review," *2022 5th International Conference on Contemporary Computing and Informatics (IC3I)*, Uttar Pradesh, India, 2022, pp. 913–920, doi: 10.1109/IC3I56241.2022.10073430.

21. "AI is the Future of Manufacturing, and It's Already Here" Website, 2023. [Online].
    Available: https://www.iiot-world.com/artificial-intelligence-ml/artificial-intelligence
    /ai-is-the-future-of-manufacturing-and-its-already-here/#:~:text=AI%20plays%20a
    %20crucial%20role,retail%2C%20communications%2C%20and%20others.
22. "Artificial Intelligence (AI) in Manufacturing Market Size, 2023" Website, 2022.
    [Online]. Available: https://www.marketresearchfuture.com/reports/artificial-intelli-
    gence-ai-in-manufacturing-market-7745

Index

Printed in the United States
by Baker & Taylor Publisher Services